2013
THE AWARD-WINNING
WORKS OF IDEA-TOPS
2013 艾特奖获奖作品集

国际空间设计大奖艾特奖组委会　编著

CSK 湖南科学技术出版社

图书在版编目(CIP)数据

2013艾特奖获奖作品集／国际空间设计大奖艾特奖组委会编著. -- 长沙：湖南科学技术出版社，2014.6
　　ISBN 978-7-5357-8147-5
　　Ⅰ.①2… Ⅱ.①国… Ⅲ.①建筑设计－作品集－世界－现代 Ⅳ.①TU206
中国版本图书馆CIP数据核字(2014)第093188号

2013 艾特奖获奖作品集

编　　著：国际空间设计大奖艾特奖组委会
责任编辑：缪峥嵘
总 策 划：深圳市东方辉煌文化传播有限公司
统筹策划：赵庆祥　梁贻攀
文字翻译：梁贻攀　夏雨娟　苏思雨
出版发行：湖南科学技术出版社
社　　址：长沙市湘雅路276号
　　　　　http://www.hnstp.com
湖南科学技术出版社天猫旗舰店网址：
　　　　　http://hnkjcbs.tmall.com
邮购联系：本社直销科 0731-84375808
印　　刷：利丰雅高印刷（深圳）有限公司
　　　　　（印装质量问题请直接与本厂联系）
厂　　址：深圳市南山区南光路1号
邮　　编：518051
出版日期：2014年6月第1版第1次
开　　本：508mm×680mm　1/16
印　　张：45.75
书　　号：ISBN 978-7-5357-8147-5
定　　价：498.00元

（版权所有 · 翻印必究）

CONTENTS 目录

6	中国境内国际化程度最高的专业设计大奖 IDEA-TOPS 艾特奖
17	2013 年度国际空间设计大奖 IDEA-TOPS 艾特奖颁奖盛典
32	2013 年度国际空间设计大奖 IDEA-TOPS 艾特奖评审委员会
36	BEST DESIGN AWARD OF CULTURAL SPACE 最佳文化空间设计奖
64	BEST DESIGN AWARD OF TRANSPORTATION SPACE 最佳交通空间设计奖
84	BEST DESIGN AWARD OF HOTEL 最佳酒店设计奖
106	BEST DESIGN AWARD OF DINING SPACE 最佳餐饮空间设计奖
134	BEST DESIGN AWARD OF ENTERTAINMENT SPACE 最佳娱乐空间设计奖
164	BEST DESIGN AWARD OF EXHIBITION SPACE 最佳展示空间设计奖
186	BEST DESIGN AWARD OF CLUB 最佳会所设计奖
210	BEST DESIGN AWARD OF OFFICE SPACE 最佳办公空间设计奖
236	BEST DESIGN AWARD OF SHOW FLAT 最佳样板房设计奖
260	BEST DESIGN AWARD OF VILLA 最佳别墅豪宅设计奖
284	BEST DESIGN AWARD OF ART DISPLAY 最佳陈设艺术设计奖
308	BEST DESIGN AWARD OF APARTMENT 最佳公寓设计奖
332	BEST DESIGN AWARD OF COMMERCIAL SPACE 最佳商业空间设计奖

中国境内
国际化程度最高的
专业设计大奖

IDEA-TOPS
艾特奖

The Highest Internationalization &
Professional Design Award in China
Idea-Tops Award

国际空间设计大奖——Idea-Tops艾特奖，是中国境内国际化程度最高的专业设计大奖，建基于全球第二大经济体及迅猛发展的设计市场，旨在发掘和表彰最佳设计师和最佳设计作品，打造全球最具思想性和影响力的设计大奖。

艾特奖为推动中西方设计交流搭建了一个沟通协作的平台，高水平的国际级评委、严谨公正的评奖机制及奖项设置……使艾特奖成为中国境内最具国际化和专业性的设计奖项，也成为世界设计业了解中国设计的一个窗口。

艾特奖参与者可谓众星云集，包括全球三大设计事务所之一的Gensler设计总监Graeme Scannell，"中国第一高塔"广州塔设计者Mark Hemel，全球酒店设计公司五强、BBG-BBGM建筑与室内设计公司设计董事Robert J.Gdowski，拥有140年历史的国际知名建筑事务所Woods Bagot全球总监Rodger Dalling，英国首相官邸——唐宁街10号设计者、BBC苏格兰总部设计师Ross Hunter，希尔顿国际酒店集团主创设计师Martin Hawthornthwaite，曼联俱乐部亚太区首席设计师Mike Atkin，深圳机场T3航站楼设计者Fuksas夫妇，2008年普利兹克奖获奖者、法国当代著名建筑师让·努维尔（Jean Nouvel）等。

源于东方，面向世界，Idea-Tops艾特奖崇尚的，是设计师永不枯竭的智慧与前瞻性的革新思想，以及他们审美和工艺上均显卓越的设计作品。恪守专业、严谨、公平、公正的原则，Idea-Tops艾特奖以较高的专业标准、专业发展、专业责任以及专业沟通来促进设计业的发展，每届艾特奖均邀请全球设计领域资深专家、学者、顶尖建筑师和设计师、知名人士、财经专家、意见领袖以及有影响力的媒体担纲评委。

作为表彰建筑和室内设计界杰出人才的重要奖项，获得艾特奖的首肯也就是向全球昭示了精英们在建筑和室内设计界的顶级荣誉。

International Space Design Award—Idea-Tops, which is the highest Internationalization & Professional Design Award in China, basing on the world's second-largest economy and the rapid development design market. The Idea-Tops award aims to discover and praise the best designers and design works, and create the most thoughtful and influential interior design award around the world.

The Idea-Tops Award, which aims to provide an exchange platform for promoting Chinese and Western design communication and collaboration. The high level of international judges, rigorous and impartial mechanism for awards and the setting of the award category, make the Idea-Tops Award become the highest international and professional design award in China, and became a window to understand China for worldwide design industry.

The Idea-Tops Award, this is a star-studded competition event, the activity participants, including Graeme Scannell, the design director of Gensler, which is one of three biggest global architectural design firms; Mark Hemel, the designer of Guangzhou Tower—China tallest tower; Robert J. Gdowski, the design director of BBG-BBGM, which is one of the top five global hotel design firms; Rodger Dalling, the global director of Woods Bagot, having over 140 history years; Ross Hunter, the designer of the No.10 Downing Street and BBC Scotland headquarters; Martin Hawthornthwaite, the chief architect of Hilton International Hotel Group; Mike Atkin, the chief designer of Manchester United plc in Asia Pacific; Massimiliano Fuksas and Doriana Fuksas, the designer of Shenzhen Baoan International Airport—Terminal 3; Jean Nouvel, France noted architect, the winner of Pritzker Prize in 2008.

The award is derived from the East, and faces to the world; it underscores the commitment to recognize designers' inexhaustible wisdom and forward-thinking innovation as well as high-quality design as defined by both aesthetic and technical expertise. With the principle: professional, rigorous, fair; the International Design Award—Idea-Tops is committed to promote the design industry by the high professional standards, professional development, professional liability and professional communication. Each session will invite a fantastic and distinctive judge panel, including senior experts, scholars, top architects & designers, celebrities, opinion leaders and influential media-men from domestic and world ranges.

As the prominent award program to recognize architectural & interior design excellence, it's said that receiving an Idea-Tops award nod is universally heralded as the top honor in the architectural & interior design industry.

郑曙旸 ZHENG SHUYANG

国务院学位委员会设计学学科评议组组长
清华大学美术学院教授、博士生导师
Idea-Tops 艾特奖评审委员会主席
Leader of Appraisal Group of Design Discipline, Academic Degree Commission of the State Council
Doctorial tutor & professor of Academy of Art & Design, Tsinghua University
Chairman of Idea-Tops Reviewing Committee

中国正处于设计文化复兴的前夜。国际空间设计大奖——Idea-Tops艾特奖的时代定位因此弥足珍贵。道路已经开通,能否走得更远,在于远大目标不懈的追求。

设计是人类基于生存的本能,以进化所达成的智慧,通过思维与表达,以预先规划的进程,按照一定的价值观,创造与之相适应的生活方式。不同生活方式历经岁月的磨砺作为历史和传统沉淀为文化。从这层意义出发,设计就是文化建设。

设计思想根植于中华民族传统文化的宏大体系,然而今日世界所认知的现代设计理论与实践,却是工业文明的产物。

设计是艺术与科学统合观念的智慧。独领风骚的人文历史与艺术传统,领先世界的科学技术传统,曾经创造了中国辉煌的古代设计文化。

华夏设计的文化传统,体现于中国人文历史的艺术价值。"中国艺术大约从公元前13世纪开始,历经沧桑,一直延续到现在。世界上没有任何一个国家能像中国那样,享有如此丰硕的艺术财富;从全面考虑,也没有任何一个国家能够与中国艺术的卓越成就相媲美。"

中国科学技术的历史地位同样显赫。中国在人类历史蜿蜒的长河中,科学技术之船在17世纪之前始终处于前列,以至英国学者李约瑟(Joseph Terence Montgomery Needham,1900—1995)在20世纪40年代发出疑问:"中国的科学技术——为什么没有朝前发展?"在学术界现在一般被称为——"李约瑟难题"。为什么在中国的古代科技领先于世界那么多世纪以后,反而是西欧于17世纪开创了现代科学技术。答案或许可以从中西文化的环境观念与形态中寻觅。

华夏文化——人与自然和谐相处的环境观:平和、舒缓、优雅、内敛;西方文化——人作为大自然主宰的环境观:极致、细密、张扬、外向。两种文化导致不同的设计理论与实践。华夏文化——统一系统的思想观念孕育农耕文明:正统、固化、维护、人文;西方文化——发散对立的思想观念引发工业文明:民权、变革、开拓、科技。

中国传统文化是超越宗教的哲学系统。道家的顺乎自然、儒家的三纲五常、墨家的一义兼爱、法家的依法治国,构建超越文明的设计伦理。2000年前的墨子,已精准阐释"衣、食、住、行"设计内涵的功能与审美。"圣人之所节俭也,小人之所淫佚也,俭节则昌,淫佚则亡。"(《墨子·辞过》)"度"之设计理念贯穿华夏千年文明。领先世界的华夏设计,体现于建筑与器型——城郭、宫室、园林、舟车、兵器;体现于材料与工艺——丝绸、陶瓷、金属、玉石、髹漆、木工、皮革、染织、刺绣、印刷……

China is on the eve of renaissance of design culture. That's why the era positioning for the international space design award—Idea-Tops seems more precious than ever. The road has been opened, but whether it can lead us to a further destination depends on our unrelenting pursuit for a more ambitious goal.

Designing is a survival-driven skill inherent in people, as well as a reflection of the evolvement of human wisdom. People design their lifestyles most matched with their sense of values based on the originally-planned schedule through ideas and presentation, and diversified lifestyles surviving as time goes by exist in the form of culture left by history and tradition. In this sense, designing is a process of culture construction.

The root of design idea can be found in the grand system of Chinese traditional culture. However, the modern design theory and practice recognized by our new world is the product of industrial civilization.

Designing is the combined wisdom of art and science. The dominating humanity history and art tradition and world-leading scientific tradition have once created the most glorious design culture in ancient China.

The cultural tradition of design in ancient China is reflected in the artistic value of China's humanity history. The history of Chinese art has gone through so many changes, starting from the 13rd century BC until now. No country in the world possesses more art resources than China does, and from a comprehensive view, no country in the world harvests more extraordinary results from art than China.

China also played a vital historical role in technology invention. Along the long course of history, China always had its scientific technology in such a leading position before the 17th century that Joseph Terence Montgomery Needham (British scholar, 1900—1995) questioned in 1940s "Why I don't see further progress in China's scientific technology?", which is now known as "Joseph's Puzzle" in academic circle. Why it was West Europe that pioneered modern technology in the 17th century after China's ancient technology had led the world for many centuries? The answer to that might be found in the environmental concept and form of Chinese and Western culture.

Ancient Chinese culture is an environment concept proposing harmony between human and nature, mild, relaxed, elegant and reserved, while western culture is an environment concept proposing human domination over nature, extreme, meticulous, conspicuous and extrovert. The two different cultures lead to different design theory and practice. Ancient Chinese culture—unified idea cultivates agricultural civilization: orthodoxy, solidification, maintenance and humanity; western culture—divergent and contradictory idea leads to industrial civilization: civil rights, reform, development and scientific technology.

Chinese traditional culture is a philosophy system beyond religion. The principles such as "Let nature take its course" proposed by Taoism, "Three guides and five virtues" by Confucianism, "Universal love" by Mohism and "Ruling by Law" by Legalism constitute the design ethics beyond civilization. Mo Tzu precisely explained the functions and aesthetic views of design for people's food, clothing, shelter and transportation. According to Mo Tzu, "those saints know better than villains that frugality leads to prosperity and extravagance to downfall". The design concept of "Degree" sustained in ancient Chinese culture for thousands of years. The leading ancient Chinese design were reflected in construction and equipment, such as castle, palace, garden, vessel & vehicle and weapon, and also in materials and crafts, such as silk, ceramics, metal, jade, lacquer coating, carpentry, leather, dyeing & weaving, embroidery and printing.

设计就是文化融会的过程，而开放则是设计发展的必需。公元前 350~ 前 220 年秦始皇统一中国的大业，奏响华夏设计文化开放的先声；东西文化交流融合的唐朝，达到开放的巅峰；经由宋代的发酵，华夏设计文化的艺术水准几近完美。公元 1421 年明朝郑和完成环球航行，郑和航海与海禁之谜，最终导致中国彻底关闭大门。即使公元 1760 年的清王朝，国土面积达到历史最大，也无法避免闭关锁国盛极而衰的命运。拒绝开放不但使国家走向衰败，华夏设计文化也未能实现从农耕文明到工业文明的转型。

当代中国亟须文化自信的重建。设计战略的时代定位：装饰的设计理念代表着传统，属于农耕文明的第一阶段；空间的设计理念代表着当代，属于工业文明的第二阶段；环境的设计理念代表着未来，属于生态文明的第三阶段。设计战术的阶段定位：体现设计本质的可持续设计——统合空间规划、构造装修、陈设装饰的整体设计；体现低碳概念的可持续设计——研究人的环境行为特征以价值体验为导向的设计；体现生态概念的可持续设计——实施创新驱动战略突破专业技术壁垒的环境设计。

中国设计从业的责任就是实现设计文化的复兴。环境设计则有可能实现率先的复兴。环境设计的传承与创新：观念层面传承系统、综合、整体的哲学观；技术层面创新以行为心理为导向的风水观。环境设计统一的时空观：以时间为导向——人为主体的设计观；以空间为导向——物为主体的设计观。环境设计应秉承时空统一以人为本的设计观。

在中国市场经济、民主政治、先进文化、和谐社会、生态文明建设的大背景下，Idea-Tops 国际空间设计大奖——艾特奖，必然顺应环境设计全球化拓展的态势，在可持续设计的全球化概念、中国传统设计文化的优势、中国室内设计的国际视野三个方面取得突破。

中国设计文化的当代性——重塑中华民族面向生态文明价值观主流文化的先进性；中国设计文化的本土性——再造国家形象自立于世界民族之林时代精神的引领性。

Designing is a process of cultural blending, while cultural opening is a necessity for the development of design. The opening and blending of the design culture in ancient China, started by the First Emperor of Qin who unified ancient China from 350 to 220 BC, reached its peak at Tang Dynasty when eastern and western cultures were mixed with each other, brewed in Song Dynasty, and finally almost came to meet the perfect artistic standards. The completion of Zheng He's worldwide voyage in A.D. 1421 of Ming Dynasty (Riddle of Zheng He's navigation and maritime prohibition) led to the end of China's opening to the world. Even the Qing Dynasty in A.D. 1760, despite its largest territory area in the history, had to succumb to its fate of downfall due to its close-door policy against the world. Seclusion not only led the country to downfall, but impeded its design culture transition from agricultural civilization to industrial civilization.

What modern China needs most is the restoration of cultural confidence. The era positioning of design strategy: design philosophy of decoration, standing for tradition, is the first stage of agricultural civilization; design philosophy of space, standing for now, is the second stage of industrial civilization; design philosophy of environment, standing for future, is the third stage of ecological civilization. The phase positioning of design strategy: sustainable design reflecting the essence of design—overall design of the overall space planning, structure decoration and furnishings decoration; sustainable design reflecting the concept of low carbon—aimed to study the characteristics of people's environmental behaviour based on value experience; sustainable design reflecting ecological concept—aimed to break through technical barriers by innovation-driven strategy.

The responsibility of Chinese designers is to realize the renaissance of design culture. Environmental design is probably the first to realize the renaissance. The inheritance and creation of environmental design: inheriting the systematic, comprehensive and integral view of philosophy at the level of conception; creating the geomantic concept which is oriented to behavioral psychology at level of technology. The unified view of time and space of environmental design: the view of design which is oriented to time and takes the human being as the subject; the view of design which is oriented to space and takes the thing as the subject. Environmental design should conform to the view of design which takes people first.

Under the circumstance of constructing market economy, democratic politics, advanced culture, harmonious society and ecological civilization, international space design award—Idea-Tops will comply with the globalization of environmental design, and obtain breakthrough on the three aspects of global conception of sustainable designing, advantages of Chinese traditional design culture and international view of Chinese interior design.

The contemporariness of Chinese design culture—rebuild Chinese nation's progressiveness facing the mainstream culture of ecologic civilization values. The locality of Chinese design culture—rebuild the guidance of time spirit of China among the nations of the world.

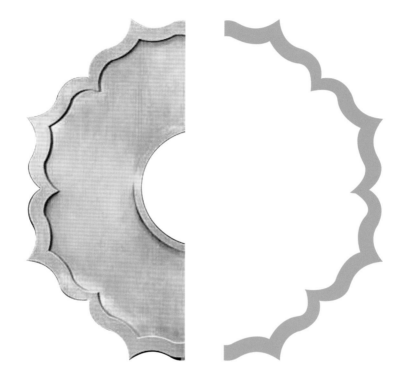

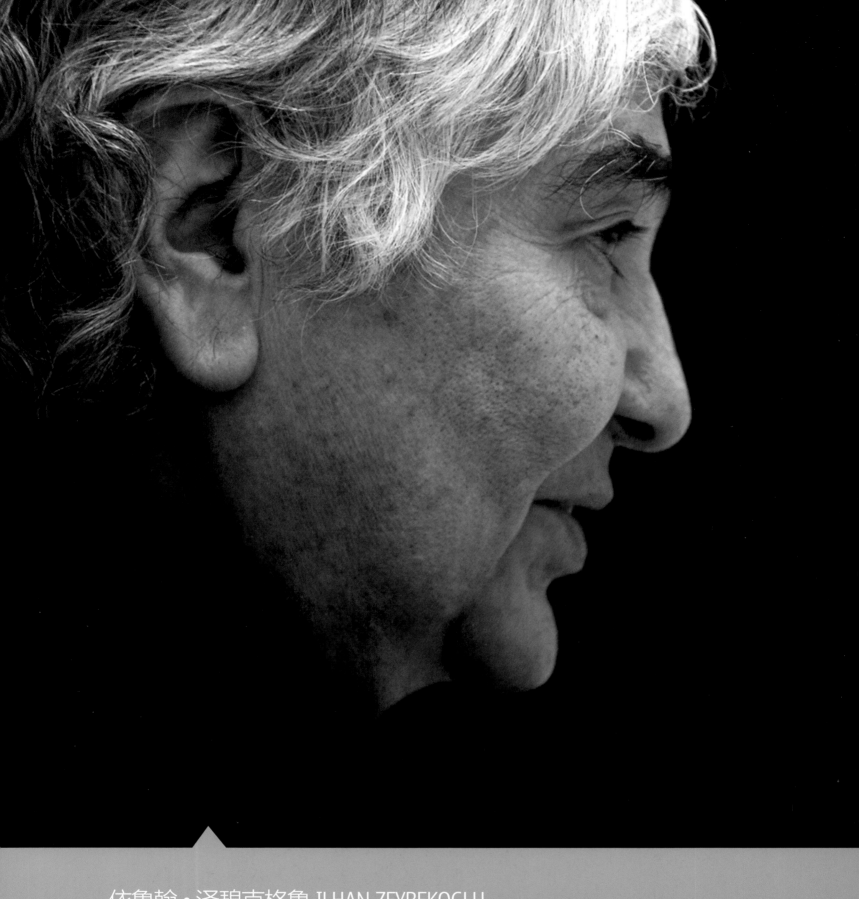

依鲁翰·泽碧克格鲁 ILHAN ZEYBEKOGLU
美国 ZNA 泽碧克建筑设计事务所创始人
Founding principal of ZNA/Zeybekoglu Nayman Associates, Inc.

我很荣幸能够作为评委参加2013年度Idea-Tops艾特奖颁奖盛典。此次参赛作品的数量和质量都相当高，这是一个鼓舞人心的信息，因为这次比赛为中国新生设计力量提供了一个展示才华的平台。这项一年一度的赛事对于中国设计产业来说是一个重要的平台，它对许多方面都有着促进作用，比如我们如何去设计、建造和使用空间。

作为今年的评委成员，我非常欣慰地看到这么多中国优秀设计师的参赛作品，以及他们对这项事业所表现出的革新与创造精神。与此同时，国外的顶尖设计师们对中国的设计行业也做出了重要的贡献。本次大赛为本土设计师提供了挑战以及宝贵的机会，使他们能够为中国市场做出非凡的贡献。他们在日常工作中形成的对文化特质的认识、洞察力与本土设计的视觉词汇一样，是他们内在拥有的东西，这对于推动和支撑当代中国的特性具有重要的意义。

设计，其最好的状态，是对经验的反映。我们时代的词汇以塑造和利用我们所生存的空间这种方式显现出来。

对于图文、韵律以及文化内涵的深刻认识，以及不同种族之间交流时刻的把握，使得设计能够满足这些需求和愿望，做到尽善尽美。

像Idea-Tops艾特奖这种比赛为年轻的设计师提供了一个场所来歌颂这种体验，并去证明设计是怎么有效表达和满足那些非常人性化的期望。

It was a pleasure and an honor to be part of this year's event, Idea-Tops Awards 2013, where I had the privilege of serving as a juror. The quality and number of entries were impressively high, a very encouraging sign, as it provided an exposure for the work of young emerging design talent from China. This annual competition serves as an important platform for the Chinese design industry and prompts a response to many aspects of how we design, build, and inhabit spaces.

As a member of this year's panel of judges, I found it was extremely gratifying to see so many entries by young Chinese designers, and the enthusiasm with which they are embracing innovation and creativity. For while leading foreign designers are also making serious contribution to design in China, competitions such as this provide both a challenge and a valuable opportunity for local designers to produce meaningful contributions to the Chinese market. Their insights and knowledge of culturally sensitive issues as well as the visual vocabulary of indigenous design are inherent in their work, and can help to drive and sustain the emerging identity of contemporary Chinese design.

Design, at its best, is a reflection of experience. The vocabulary of our days becomes manifest in the way we shape and use the spaces we inhabit.

A heightened awareness of the patterns, rhythms, and textures of culture and the moments that punctuate different human interactions allow design to serve those needs and desires responsibly and aesthetically.

Competitions such as Idea-Tops Awards, offers young designers a forum to celebrate those experiences, and to demonstrate how design can effectively reflect and satisfy those very human expectations.

赵庆祥 ZHAO QINGXIANG
艾特奖组委会执行主席
Executive chairman of Idea-Tops Award Organizing Committee

今年，艾特奖即将进入第五届了。作为艾特奖的发起人，我深感荣幸，艾特奖为地产界、酒店投资及管理集团、更多的商业投资项目与当今顶尖设计师合作交流、项目对接提供了绝佳平台，也正是基于艾特奖的影响力、公信力和平台作用，能够让来自世界各地最优秀的设计师在每年的12月相聚在一起。

发掘和表彰在技术应用、艺术表现及文化特质再现上，具有创新意识的设计师和设计作品，打造全球最具思想性和影响力的设计大奖，以最高的专业标准、严谨的专业责任以及良好的专业沟通来促进设计业的发展，这是艾特奖的使命，正是坚守了这一使命，艾特奖专业影响力、学术影响力、国际影响力日益凸显，并从众多的设计奖项、赛事中脱颖而出，受到来自世界范围越来越多的关注。

过去几届，艾特奖在意大利米兰、美国洛杉矶、北京、上海、深圳、广州、沈阳、杭州等20多个城市开展了巡回推广活动。特别是去年9月，艾特奖在美国洛杉矶成功举办了中美设计峰会和艾特奖启动仪式，凤凰卫视美洲台以及美国近10家主流媒体对艾特奖进行了系列报道，将艾特奖的影响力传播至海外。

当今中国已成为全球第二大经济体，城市化建设和市场的迅猛发展，为设计师提供了广阔的施展才华的空间。更多欧美等发达国家的设计师希望到中国来开辟自己的设计市场，艾特奖的国际参与度也因此越来越高。2010年，艾特奖参赛设计作品616件，2011年1596件，2012年2616件，2013年共收到来自35个国家的参赛设计作品3661件，每年递增上千件。

我们高兴地看到，在2013年度艾特奖的颁奖礼上，上千名现场观众在各大奖项揭晓时发自内心的热烈掌声；在中央电视台等国家级媒体，对于艾特奖专业性的客观报道和对获奖设计作品的高度评价。国际评委们也对获奖作品给予了高度肯定，"它们在一定程度上具有国际领先和示范作用，比如深圳机场T3航站楼可以称得上是地标性的建筑。"我们也十分乐意见到，越来越多的房地产开发商和酒店投资方，以艾特奖获奖设计师作为优先选择的合作伙伴。

今天，艾特奖已经成为诞生明星设计师的舞台，发现有思想、有才华的杰出设计师，用聚光灯和标杆的力量推动整个行业的进步是艾特奖的任务。伴随着中国城镇化建设的进一步加快，伴随着中国在世界舞台地位的提升，相信透过艾特奖这一平台，一定会诞生属于中国本土的设计大师，我期待着这一天的到来！

This year,the Idea-Tops Award is marching into the fifth year. As a sponsor of Idea-Tops Award, I am deeply honored, the Idea-Tops Award provides an excellent platform for the real estate sector, hotel investment and management group, more business investment cooperation and exchanges with today's top designers, project docking, also the influence, credibility and a platform are based on Idea-Tops Award allow the best designers from around the world come together each year in December.

The mission of the Idea-Tops Award are to explore and recognize in the technology, artistic expression and cultural characteristics reproducing, with innovative designers and design work to create the world's most thoughtful design awards and influence to the highest professional standards, rigorous professional responsibility and good to promote the development of professional communication design industry, due to hold out this mission ,professional influence, academic influence and international influence have become increasingly prominent,the Idea-Tops Award stand out from the numerous design awards ,get attention from around the world more and more.

Past sessions, the Idea-Tops Award have launched a campaign tour in Milan, Los Angeles, Beijing, Shanghai, Shenzhen, Guangzhou, Shenyang, Hangzhou ,etc,over 20 cities. Especially last September, the Idea-Tops Award successfully held summit the China-US and Ait design awards ceremony started in Los Angeles, Phoenix North America and the United States about ten mainstream media carried a series of reports .The influence of the Idea-Tops Award will spread overseas.

Nowadays China has become the world's second largest economy, for designers, urbanization and the rapid development of the market provide a broad space to display their talent. More designers from Europe and America and other developed countries hope to come to China to open up their own blueprints, therefore the international participation of the Idea-Tops Award is increasing.In 2010, there are 616 competition designs,in 2011, there are 1596 designs, in 2012 there are 2616 designs,in 2013 3661 designs were received from 35 countries ,more than thousands designs is increasing every year.

In 2013 the Idea-Tops Award, we are pleased to see that, thousands of spectators clap hardly from heart when the major awards announced; the professional of the Idea-Tops Award was reported and award-winning design is highly evaluated by CCTV and other national media. International judges also give highly positive to award-winning design, "they have an international leader and model role, such as Shenzhen Airport T3 can be called a landmark building." We are also very pleased to see that, a growing number of real estate developers and hotel investors select partners from award-winning designers.

At present, the Idea-Tops Award is a stage where many leading designers are emerging. The Idea-Tops Award which find designers with thought and talent promote the progress of the entire industry with the power of the spotlight and benchmarking. With further accelerate China's urbanization, along with China in enhancing the status of the world stage, we believe that through this platform Ait Award,a outstanding designers will be born in the Chinese mainland,I look forward to the arrival of this day!

2013年度国际空间设计大奖
IDEA-TOPS 艾特奖 颁奖盛典

International Space Design Award
Idea-Tops 2013
Awards Ceremony

12月8日，2013年度艾特奖颁奖盛典在深圳保利剧院举行。来自美国、法国、德国、意大利、希腊、西班牙以及中国大陆20多个城市的1400多名设计师代表出席本届艾特奖颁奖盛典活动，台湾、香港、澳门以及中国大陆近20个城市的设计师均组团参加。

最终，2008年普利兹克奖获得者、法国当代著名建筑师让·努维尔的设计工作室，与深圳机场T3航站楼的设计者——意大利福克萨斯公司，以及来自德国、葡萄牙、希腊、比利时、西班牙和中国台湾、重庆、厦门、宁波、杭州等的设计师分获本届艾特奖的13项大奖。

中央电视台CCTV-1、CCTV-13（新闻频道）《新闻直播间》、香港卫视、深圳卫视、南方卫视、广东电视台、凤凰网、中国网、《南方日报》、《深圳特区报》、《深圳商报》、《深圳晚报》、《广州日报》等近百家主流媒体对本届艾特奖进行了报道，在社会各界引起广泛关注和强烈反响。12月9日，中央电视台以"艾特设计奖首发，定位城市文化"为题报道本届艾特奖，并评价"艾特奖是国内国际化程度最高的专业设计奖项"。

The Idea-Tops Award 2013 ceremony was held in Shenzhen Poly Theater on December 8. There were 1400 designer representatives, who came from the United States, France, Germany, Italy, Greece, Spain and more than 20 cities of mainland China, attended the awards ceremony. Meanwhile there were nearly 20 cities from Taiwan, Hong Kong, Macau and mainland China designer representatives formed teams to participate in the award ceremony event.

Finally, the contemporary French design studio—Atelier Jean Nouvel, the winner of Pritzker Prize in 2008, the Italy FUKSAS, the designer of Shenzhen Baoan International Airport—Terminal 3 and other designers from Germany, Portugal, Greece, Belgium, Spain, Taiwan, Chongqing, Xiamen, Ningbo and Hangzhou won the final Idea-Tops Award for 13 different categories.

CCTV-1, CCTV-13 *Live News*, HKSTV, Shenzhen TV, TVS, Guangdong TV, ifeng.com, China.com, *Nanfang Daily*, *Shenzhen Special Zone Daily*, *Shenzhen Economic Daily*, *Shenzhen Evening News*, *Guangzhou Daily* and other hundreds of mainstream Medias reported the award ceremony, attracting widespread concern and strong repercussions from all circles of life. CCTV made a report, which entitle as the Idea-Tops Award goes its ways, and positions urban culture, and evaluated the award as the highest internationalization & professional design award in China.

LOS ANGELES

2013年9月，艾特奖组委会在美国洛杉矶举办了艾特奖的境外推广活动，中国驻洛杉矶总领馆领事高洪善先生出席活动并接受各大境外媒体采访，凤凰卫视美洲台对艾特奖进行了系列报道，将艾特奖的影响力传播至海外。

The Idea-Tops Award Organizing Committee held overseas promotion activities in Los Angeles on September 2013, Mr. Gao Hongshan, the consular officer of the Consulate-general of the People's Republic of China in Los Angeles, attended the event and accepted different major foreign media interviews, Phoenix Satellite Television (US) Inc held a series of reports about the event, spreading the overseas influence of Idea-Tops Award.

"移植现代"主题对话一:"文化穿越"
--中华文化与当代设计高峰对话

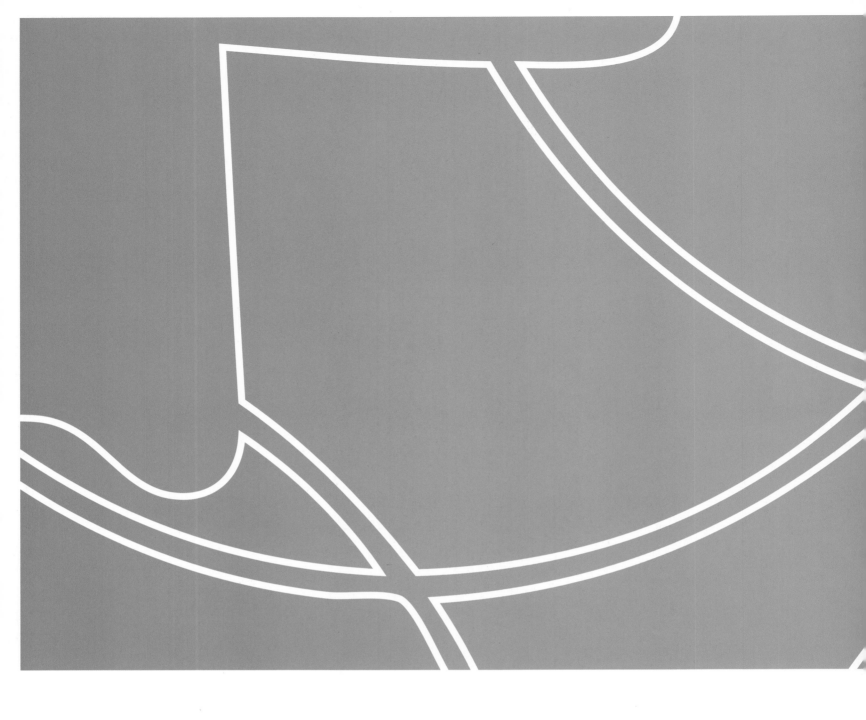

2013 年度国际空间设计大奖 IDEA-TOPS 艾特奖
International Space Design Award—Idea-Tops 2013

评审委员会
Evaluation Committee

¹ 郑曙旸 ² ILHAN ZEYBEKOGLU
³ JEAN-PAUL CASSULO ⁴ 赵红红 ⁵ 梁景华
⁶ 张月 ⁷ 王铁 ⁸ 彭军 ⁹ 马克辛
¹⁰ 龚书章 ¹¹ 史迪威 ¹² 潘鸿彬

_____ 郑曙旸

国务院学位委员会设计学学科评议组组长、清华大学美术学院教授、博士生导师。中国建筑学会室内设计分会副理事长、中国室内装饰协会常务理事、设计委员会副主任，中国美术家协会环境艺术设计专业委员会副主任，中国建筑装饰协会信息咨询委员会委员。主要设计项目：驻德国大使馆室内设计、中南海紫光阁国务院接见厅室内设计、中国丝绸进出口总公司办公楼室内设计、中国远洋运输总公司办公楼室内设计、敦煌宾馆贵宾楼室内设计、国务院接待楼室内设计、首都国际机场航站楼室内设计等。

_____ Ilhan Zeybekoglu

美国 ZNA 建筑事务所总裁、前美国哈佛大学建筑学院教授、美国注册建筑师协会专家。泽碧克格鲁先生，担任此次 AIM 国际设计竞赛评委会主席，美国 ZNA 建筑事务所的主要创始人，资深建筑师，前哈佛大学教授，波士顿建筑学会元老。80 年代创设 ZNA 事务所之前曾师从贝聿铭，至今建筑规划专业设计经验超过 40 年，获奖无数，载誉业界。曾参与许多国际规划及建筑设计项目，屡获殊荣。例如开罗美国大学整体规划，沙特阿拉伯卡利德国王大学校园建筑群、迪拜健康产业城（DHCC）、中国哈西新区大楼、京哈高铁新客站站前广场等，都是近年来的出彩项目。泽碧克格鲁先生是沙特阿瓦利德王子的规划开发顾问团成员之一，他也是美国联邦政府、沙特阿拉伯王国政府、阿联酋王室等知名商业、政府机构的常年设计顾问。

_____ Jean-Paul CASSULO

国家文凭建筑师、法国国家建筑师行会"普罗旺斯－阿尔卑斯－蓝色海岸"行政大区主席、法国建筑、城市规划及环境顾问委员会联合会建筑师委员、法国"交通与住宅部"颁发"银绶带"奖获得者（1997年）(注"普罗旺斯－阿尔卑斯－蓝色海岸"行政大区位于法国南部，毗邻地中海。包含了6个省：上普罗旺斯阿尔卑斯省（04）、上阿尔卑斯省（05）、阿尔卑斯滨海省（06）、隆河河口省（13）、瓦尔省（83）及沃克吕兹省（84）。大区首府：马赛市。

_____ 赵红红

国务院政府特殊津贴专家，华南理工大学广州学院建筑学院院长，博士生导师，国家一级注册建筑师、国家注册城市规划师，华南理工大学城市规划与环境设计研究所所长，兼任中国城市规划学会第三届理事会理事、中国建筑学会第十一届理事会理事、全国高等学校建筑学专业指导委员会副主任委员、全国风景园林硕士专业学位教育指导委员会委员、广州市城市规划协会常务理事、副理事长、广州市人民政府决策咨询专家、广州市城市规划委员会委员、佛山市顺德区城市规划委员会委员、广州市城市景观协会第一届理事会副会长、广东省房地产协会专家咨询委员会副主任委员等。

_____ 梁景华

香港十大设计师，PAL 设计事务所有限公司创始人、首席设计师，英国特许设计师协会会员、澳洲设计学会国际会员、香港室内设计协会荣誉顾问／前副主席。2010 年度 Idea-Tops 艾特奖"最佳会所设计大奖"获得者。1999 年获亚太区室内设计大奖会所组冠军；2001 年再次获得亚太区室内设计大奖会所组冠军；2005 年荣获亚太区室内设计大奖商业组金奖；2006 年获第五届中国企业创新人物表彰大会颁发的"中国企业创新优秀人物"荣誉称号，并在同年当选香港 2006 年十大杰出设计师。

_____ 张 月

清华大学美术学院环境艺术设计系主任，教授，硕士生导师，中国室内装饰协会常务理事，中国室内装饰协会注册设计师评委会评审委员，中国工业设计协会会员，教育部高等教育文科计算机基础教学艺术类分委会委员，北京市高等教育自学考试办公室环境艺术设计类专业委员和课程委员。主要设计项目包括阿拉伯也门共和国马里普饭店室内设计、北京人民大会堂甘肃厅室内设计、北京中南海怀仁堂中共中央会议厅室内设计、中央军委办公大厦室内设计、北京中国国际贸易中心室内设计、中国驻纽约领事馆室内设计等。所获奖项有首届中国室内设计大展金奖、第二届中国室内设计大展银奖、北京市高等教育教学成果一等奖、2006 年中国室内设计双年展获室内设计突出成就奖等。

---------- 王 铁

中央美术学院学术委员会委员、教授、硕士生导师、特级景观设计师，深圳市室内设计师协会高级顾问、青岛完美家居研究所所长、中国建筑装饰协会专家组副组长。曾参与钓鱼台国宾馆12号楼四季大厅设计、清华大学美术学院（原中央工艺美术学院）教学主楼建筑设计、中国银行国际金融研修院室内设计、日本名古屋富士町综合商业大楼建筑设计、日本名古屋内山集合住宅建筑设计等。发表论文多篇，著有《室内设计与环境》等。

---------- 彭 军

天津美术学院艺术设计学院副院长、教授。英国布鲁乃尔大学、诺森比亚大学高级访问学者，全国有成就的资深室内设计师、中国美术家协会会员；中国建筑装饰协会设计委员会副主任、中国室内装饰协会设计委员会副秘书长、全国高等美术院校建筑与环境艺术教育年会组委会秘书长；天津美学学会副会长、天津市照明学会副理事长、天津环境装饰协会常务理事、天津市容委夜景灯光专家委员会成员、天津市建筑工程评标专家。

---------- 马克辛

鲁迅美术学院环境艺术设计系主任、教授、硕士研究生导师。联合国教科文卫专家组大中华地区专家、国际室内装饰设计协会IFDA中国理事、国际商业美术师协会ICADA特级设计师、中国美术家协会环境艺术委员会副主任、中国建筑装饰协会设计委员会副主任、中国美术家协会壁画艺术委员会理事、中国流行色设计协会理事、中国室内装饰协会设计委员会副主任、中国城市雕塑专家评审委员会委员、辽宁省教育厅高级职称评审专家，第十届、第十一届全国美展评委、终评委，2010年上海世界博览会景观设计顾问专家。

---------- 龚书章

中国台湾著名建筑设计师，现任中国台湾交通大学建筑研究所副教授兼所长，曾任中国台湾原相联合建筑师事务所主持建筑师。哈佛大学设计硕士、建筑硕士、中国《台湾建筑》杂志编辑委员、台北市建筑师公会登记建筑师、高雄市建筑师公会登记建筑师、中国台湾建筑师公会登记建筑师。

---------- 史迪威

中国台湾著名设计师，美国哈佛大学设计硕士，上海元柏建筑设计事务所负责人，中国台湾中国文化大学建筑学学士，美国麻省理工学院助理讲师，上海交通大学客座教授，曾在美知名的建筑师事务所（Jung/BrannenAssoc.Inc.）工作7年之久，长年在美国与中国台湾著名建筑设计公司主持建筑与室内的设计规划业务，现任伦达建筑设计有限公司副总经理。他的作品跨越建筑及室内领域。

---------- 潘鸿彬

香港理工大学设计学院室内设计学荣誉学士及设计学硕士，PANORAMA泛纳设计事务所创始人。香港理工大学设计学院助理教授，香港室内设计协会副会长，IFI国际室内建筑/设计师联盟执委。曾获殊荣：包括美国IDA设计大奖、日本JCD设计大赏100强、荷兰FRAME Great Indoors Awards提名、iF中国设计大奖、中国最成功设计大奖、金指环–iC@ward全球室内设计大奖、APIDA亚太室内设计大奖、HKDA香港设计师协会环球设计大奖、PDRA透视室内设计大赏及入围DFA亚洲最具影响力设计大奖、2008/2010年度香港十大杰出设计师大奖、中国优秀创新企业家等，2011年中国台湾国际室内设计大展被邀参展的10位国际设计师中唯一华人。

最佳**文化空间**设计奖
BEST DESIGN AWARD OF
CULTURAL SPACE

最佳文化空间设计艾特奖 Best Design Idea-Tops Award of Cultural Space
Ateliers Jean Nouvel, Laurent Duport, Nicolas Cregut

最佳文化空间设计提名奖 The Nomination for Best Design Award of Cultural Space

郭海兵(中国上海)/ Ippolito Fleitz Group GmbH(德国)/ 兰敏华(中国深圳)/ Philippe CERVANTES, Philippe BONON, Gilles GAL(法国)

艾特奖

最佳文化空间设计

Best Design Idea-Tops Award of Cultural Space

RBC Design Center Montpellier
蒙彼利埃宏利保险公司设计中心

设计者 | Designer

Ateliers Jean Nouvel, Laurent Duport, Nicolas Cregut（法国）

设计说明 | Design Illustration

这座由让·努维尔设计的建筑可以首先被视为一个简洁的盒子。一旦进入此建筑，室外一切皆通透可见。此建筑深刻独特的灰色基调就如同是件装配好的家具，是辉煌雄伟的博物馆的一个完美的场景。中央深井的两边分别分布了8个平台，被不锈钢针支撑保护着，它们构成了此空间内的颜色和生命，邀请您进行一场接一场达到完美要求的视觉漫步之旅。这壮观的"家具衣柜"，由阶梯连接各级平台，像是受到了数学家埃舍尔的画的启发。此建筑位于玛丽安港口，蒙彼利埃一个全新的当代的区域位置，这里的市政厅同样也是由努维尔设计的。

The building designed by Jean Nouvel can be first seen as a simple box. Once you enter, the outside opacity gives place to transparency. The deep unique grey tone of the building is a perfect scene for the strong museum like set ups of the furniture pieces. Distributed on eight levels on both sides of a central major hole, protected by a stainless stitch, they are colors and life of the place, inviting to roam from a visual request to the other one. This spectacular "furniture wardrobe", cut by stairs that link levels, looks inspired by mathematician Escher's drawings. The building is situated in Port Marianne, a brand new contemporary area in Montpellier where Jean Nouvel has also designed the town hall.

一层层的平面设计，白色的线条与透明的玻璃交相叠加，就像一块千层糕。这种设计将建筑物的比例抽象化，使物品看上去既可以像一件家具，又有可能是城市里某个建筑物的缩影。简单的材料，真实的光线，在城市里形成了一个既多变又与周围环境完美融合的建筑。同时，它的外观和室内设计都极具现代感，像一个正待开启的珠宝匣，白天明亮，夜晚魔幻。总之，这是一次非常美妙的建筑设计之旅。

——Jean-Paul CASSULO 让·保罗·卡苏洛

（法国国家建筑师行会"普罗旺斯-阿尔卑斯-蓝色海岸"行政大区主席）

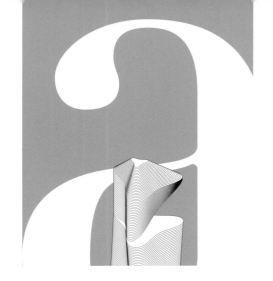

提名奖

最佳文化空间设计

The Nomination for Best Design Award of Cultural Space

Heihe City Planning Experience Hall
黑河市城市规划体验馆

设计者 | Designer

郭海兵 Guo Haibing

设计说明 | Design Illustration

充分利用具有特色艺术与历史符号的老厂房建筑优势，并考虑到展示与建筑在表达语汇上的一致性，在保留建筑结构的同时对内部空间进行优化调整。合理设计参观动线，力求从空间布局、艺术表现、材质选择等各方面呈现出将当代设计与老建筑文脉相融合的特质，做到一脉相承、内外兼修。

Making full use of the advantages of the old factory building distinctive symbol of art and history,taking into account the consistency in presentation and architectural vocabulary of expression,retaining structures while optimizing the adjustment of the internal space,designing the visiting moving lines rationally,sought from the spatial layout, artistic expression, material selection and other aspects to showing the characteristics of contemporary design and the integration of the old architectural context,to makes the design same strain,internally and externally.

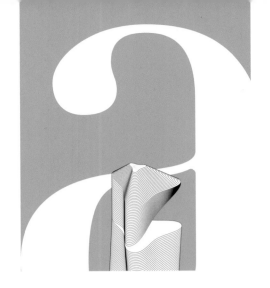

提名奖

最佳文化空间设计

The Nomination for Best Design Award of Cultural Space

Schorndorf Cityhall

绍恩多夫市政厅

设计者 | Designer

Ippolito Fleitz Group GmbH

设计说明 | Design Illustration

这座建于1726至1730年的市政厅是市中心广场上唯一的建筑，也是绍恩多夫市（Schorndorf）的象征。在第一期的改造工程中，首先对二层以上的部分从节能和文物保护的角度进行了修缮。第二期工程则对一层的公共空间以及外墙进行了重新设计。这一设计的主旨是在最大程度上展现这座建筑的历史空间。对所有区域的透明处理象征了贴近公民的民主意识，并且让人回想起底层原本作为公共广场的功能。从矗立着一根根古老木柱的一层大厅顺台阶向上可前往婚礼厅。婚礼厅漂浮般的结构和下面议事大厅均采用玻璃包裹，令整个空间保持了整体性。议事大厅风帆状的天花板设计令人印象深刻，座位设置有利于目光交流，促进了与会者之间建设性的交流气氛。

Originally built in 1726—1730, Schorndorf cityhall is the jewel and landmark of the town of Schorndorf. In the first stage of construction, we energetically rehabilitated the top floors of the building and restored them in line with historical monument regulations. A second phase of construction followed with a redesign of the public area of the ground floor and a new facade design. The main idea behind the design was to make the historical space as visible as possible. The resulting transparency in all areas demonstrates a citizen-oriented understanding of democracy and recalls the former marketplace function of the ground floor.

A staircase leads up from the foyer with its historical wooden pillars to a glass-fronted wedding hall. This free-floating unit with a subjacent window to the plenary hall are made of glass, so the room can be perceived as a whole. The plenary hall is characterised by an undulating ceiling sail and a seating arrangement that is optimised for maximum eye contact, thus promoting a constructive atmosphere for communication.

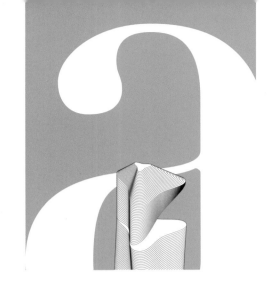

提名奖

最佳文化空间设计
The Nomination for Best Design Award of Cultural Space

Shenzhen Library—South Bookroom
深圳图书馆—南书房

设计者 | Designer

兰敏华 Lan Minhua

设计说明 | Design Illustration

设计师巧妙地将深圳图书馆的经典馆藏、特殊文献展示和互动分享的读书体验结合在一起，运用朴实的色彩与造型在错落的空间中意图表达"道·法·自然"的空间概念。这一概念倡导万事万物应遵循"无生有，有归无"的宇宙法则，因为它是自然的起点，也是运动变化的根本依据。

Designers skillfully combine the classic collection of Shenzhen Library with display and interactive sharing experience of reading special literature together, using simple intention expression in color and modeling in the space of strewn at random "tao, law, natural" spatial concept. The concept of advocating everything should follow "everything start from nothing, and stop in it" the laws of the universe, because it is the starting point of the nature, also the fundamental basis of movement.

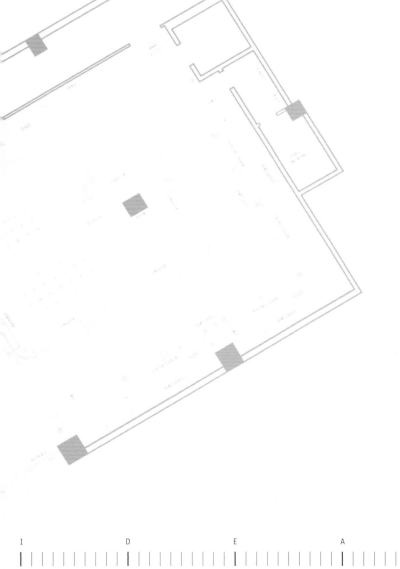

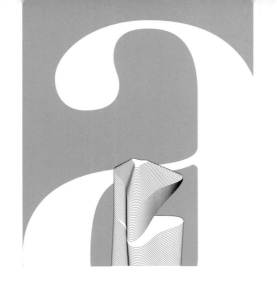

提名奖

最佳文化空间设计

The Nomination for Best Design Award of Cultural Space

Theater Jean-Claude Carriere

设计者 | Designer

Philippe CERVANTES, Philippe BONON, Gilles GAL

设计说明 | Design Illustration

建筑物采用红色作为主体颜色，给人热情活泼的视觉感受，很符合剧院的风格，外框是木材组建的多种菱形，与玻璃的菱形框架相呼应，整体达于和谐。

Red as the main body color of the building, which brings people warm and lively feelings, that is up to the style of theatre, wood forms a variety of rhombus outer frame, which is in accordance with rhombus glass framework, achieving a harmonious entirety.

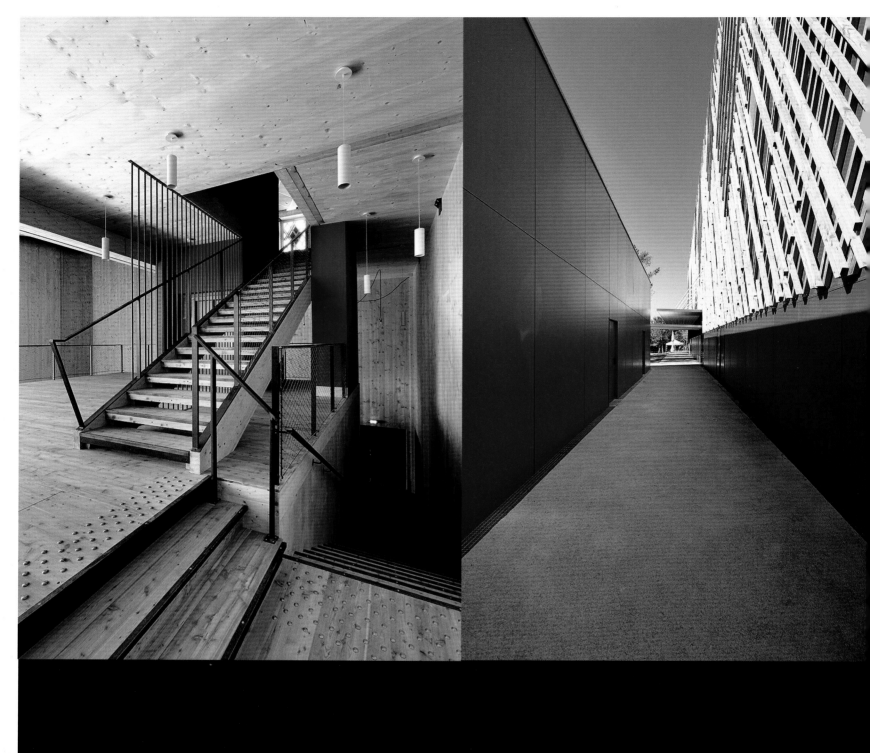

最佳**交通空间**设计奖
BEST DESIGN AWARD OF
TRANSPORTATION SPACE

最佳交通空间设计艾特奖 Best Design Idea–Tops Award of Transportation Space
福克萨斯建筑设计事务所 FUKSAS

最佳交通空间设计提名奖 The Nomination for Best Design Award of Transportation Space

HENN Architect / 吕氏国际室内建筑事务所 / 中旭建筑设计有限责任公司理想空间工作室

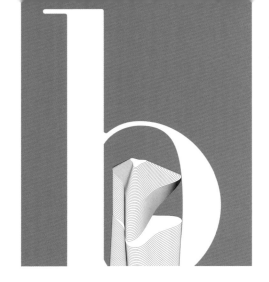

艾特奖

最佳交通空间设计
Best Design Idea-Tops
Award of Transportation Space

Shenzhen Baoan International Airport - Terminal 3
深圳宝安机场 T3 航站楼

设计者 | Designer

福克萨斯建筑设计事务所 FUKSAS

设计说明 | Design Illustration

深圳宝安机场 T3 航站楼融合了国际先进的设计理念，其优雅而独特的"飞鱼"造型，向世人展示出一个激动人心且具有时代感的建筑。航站楼钢结构设计新颖，在制作及施工方面独树一帜。实现了形式与功能的完美结合，除了自然采光外，还造就了变幻无穷的光影空间。形状大小各异的天花板呈现出的立体三维效果，使人仿佛沐浴在一片缤纷海洋与光彩照耀之中。航站楼幕墙玻璃采用六边形造型，通过充分利用自然光源保证航站楼内的光线充足，全年可以减少照明用电 470 万度。出发大厅内天花均为立体镂空铺设，能最大限度地利用自然光线；西侧的天花板呈平面状，能有效阻挡从西边射过来的太阳光，使候机楼内更为清爽宜人，减少空调使用量，达到节能的目的。

Shenzhen Baoan International Airport - Terminal 3 combines international advanced design concept,its elegant and unique "flying fish" shape,demonstrates an exciting contemporary architecture to the world. In terms of the mature technique of production and construction and the unique steel design achieve the perfect combination of form and function.The designers deal well with the natural light problem,and create a space,with countless changes shadow. The ceilings, different shapes and sizes show a stereoscopic three-dimensional effect;it seems that you are bathed in a glory shine and vast ocean. The screen wall is designed as hexagonal shape,to make full use of natural light to ensure adequate lighting in the interior space of terminal;this measure can save 4.7 million degrees electricity for lighting in one year.The ceiling of departure hall are designed in three-dimensional hollow shape, which can maximize the use of natural light;the ceiling of the west side can effectively block the shot comes from the west of the sun,which can make the interior terminal much more refresh,and reduce the using of the air-condition, achieving energy saving purpose.

A brilliant expression of public space, exploring the clasity of movement, way finding and lighting.
公共空间丰富的设计表达，动线的展示，轨道设计以及灯光设计都是本次设计的出彩之处。

——Ilhan Zeybekoglu 依鲁翰·泽碧克格鲁
（美国 ZNA 泽碧克建筑设计事务所创始人、美国哈佛大学建筑学院教授）

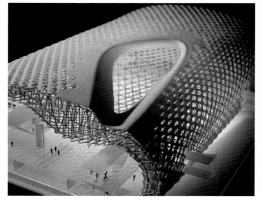
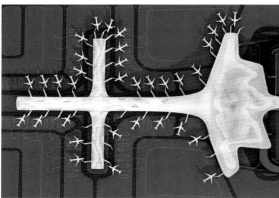

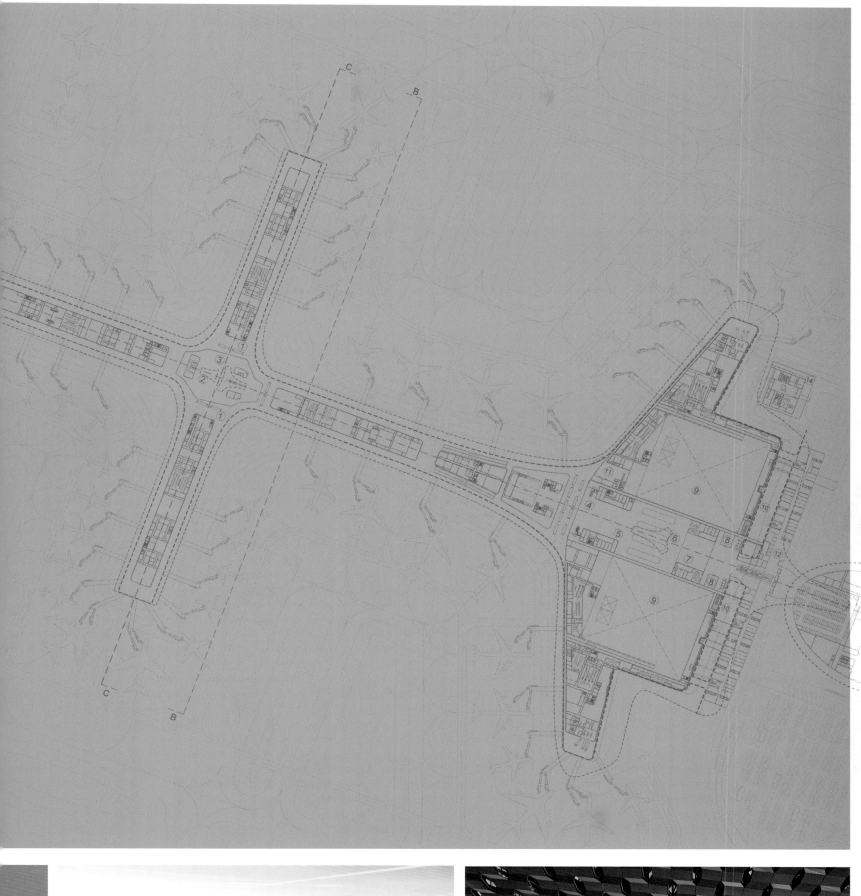

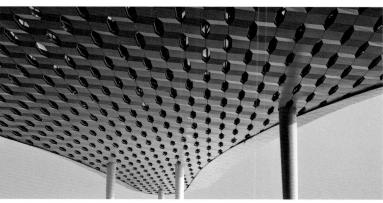

提名奖

最佳交通空间设计
The Nomination for Best Design Award of Transportation Space

VW AutoTurme

设计者 | Designer

HENN Architect

设计说明 | Design Illustration

沃尔夫斯堡大众汽车城的北面是一片长方形的水域，东北面有两个玻璃圆柱体，那就是透明汽车塔。这两座塔象征着汽车的生产流程，并把即将交货的新车储存起来。每座塔有20层，能容纳400辆汽车。机械手臂型的电梯将刚刚驶下生产线的新车运送到20层停车楼中空闲的停车位。每40秒就有一辆新车到从生产车间抵达圆柱透明塔楼的地下，同时将有另一辆车朝交付中心开离。圆柱透明塔楼内部稳定的运行就像是一个玻璃马达，制造出沃尔夫斯堡汽车城的有力心跳。

The northern limitation of the Autostadt Wolfsburg is created by a long, rectangular water basin with two glass cylinders to the northeast: the Auto Turme. They all symbolize the process of vehicle production and serve as an intermediate storage for the newly completed cars prior to delivery. Every tower holds 400 vehicles which are transported via elevators into the free shafts of the 20 storey. Every 40 seconds a new car arrives underground at the cylinder from the production plant, while another leaves the towers in the direction of the Kunden Centre. This constant movement to the inside of the cylinder as a glass motor sets the heartbeat of the Autostadt Wolfsburg.

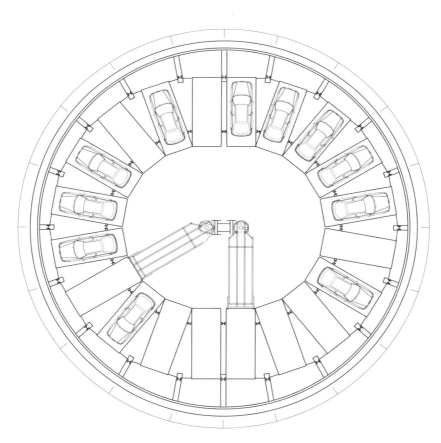

提名奖
最佳交通空间设计
The Nomination for Best Design Award of Transportation Space

Shenzhen North Railway Station
深圳北站综合交通枢纽

设计者 | Designer

LVSHI International Interior Architect Office
吕氏国际室内建筑事务所

设计说明 | Design Illustration

深圳北站主体为大跨度异型钢结构设计，采用国内外首创的结构体系。屋盖为双向双层不规则截面桁架钢结构，下部为钢管砼柱与钢砼楼板组合梁框架结构。车站分地面站台层、高架夹层、高架候车层、商务预留层共四层。深圳北站建筑结构形式美观、新颖，大量新材料、新工艺在国内尚属首次，已成为深圳标志性建筑。

The main building of Shenzhen North Station is a large-span steel structure,taking advanced technique for architecture system.The roof of the building is an irregular steel structure,with cross-section and two-way,the concrete columns and steel reinforced concrete slab composite frame structure of the lower part.The station contains the ground platform layer,elevated mezzanine layer;elevated waiting layer and business layer four stories.The building structure of Shenzhen North Station is beautiful and innovative;many new materials and new technology are used as the first time in China,Shenzhen North Station is one of the landmarks in Shenzhen.

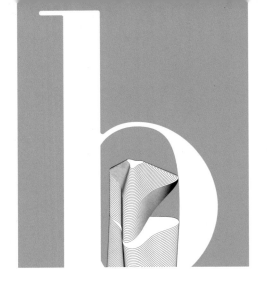

提名奖

最佳交通空间设计

The Nomination for Best Design Award of Transportation Space

Erdos Airport
鄂尔多斯机场

设计者 | Designer

Zhongxu Architecture Design Co.,Ltd .Idea-Space
**中旭建筑设计有限责任公司
理想空间工作室**

设计说明 | Design Illustration

室内设计力求干净、明快。采用模数化、拼装化，强调视觉通透性和使用舒适性，将现代语言与民族及地域特点结合起来。创造性的设计解决了多维曲面的平滑安装，巨型天顶画的安装及吊装，放射性天花板的模数化，高寒地区室内种植，多功能综合标识系统、多功能台面等一系列问题，并将机场远期使用中必将不断变化的广告及商业需求进行了预留考虑。

The designer creates a clean and bright interior space.The designer emphasizes on the sense of the visual permeability and the use of comfort by modular and assembled way,and combines modern languages and ethnic and geographic characteristics abstractly.The creative design,installation of multi-dimensional in the smooth surfaces,installation and lifting of the huge ceiling paintings,the radioactive ceiling modulus, indoor cultivation in the alpine region,multi-functional integrated identification system, multi-function table and other issues make such a successful project.Meanwhile the designer takes consider of the long-term use of airport advertising and changing business requirements.

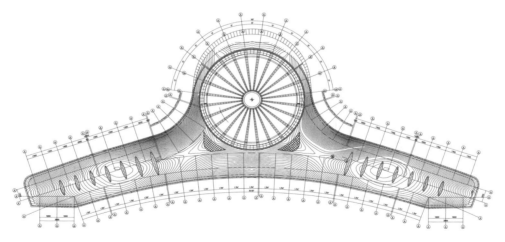

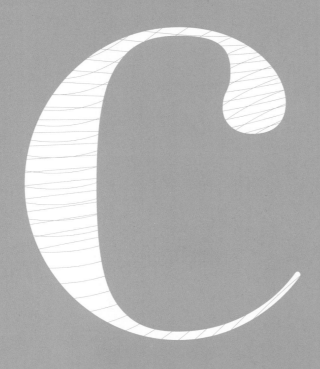

最佳**酒店**设计奖
BEST DESIGN AWARD
OF HOTEL

最佳酒店设计艾特奖 Best Design Idea–Tops Award of Hotel
D Zege Architects

最佳酒店设计提名奖 The Nomination for Best Design Award of Hotel
杨邦胜酒店设计顾问公司 / 深圳市同心同盟装饰设计有限公司 / 纳杰

艾特奖

最佳酒店设计
Best Design Idea-Tops
Award of Hotel

The Met Hotel

设计者 | Designer

D Zege Architects

设计说明 | Design Illustration

The Met Hotel 是一家五星级酒店，该项目由 Zege 建筑事务所为 Chandris 集团设计完成，该项目已然成为一个新的城市地标性建筑，吸引了大众的眼球，这样的影响力对于一个位于前工业环境区的建筑是前所未有的。米特酒店也有助于附近港口的发展，也是建筑能带给城市新生命很好的演绎。项目的整体设计让酒店、大海以及码头有一个很好的融合，通过一些特殊的处理方式，让大海、码头的元素融入酒店设计中，波浪形的外墙设计，黑白相间的大理石，酒店整体布局以及酒吧锯齿线设计等。前台和餐厅给人一种运动感，让直线突破空间。房间内大窗户以及浴室透明墙的设计，让酒店客人对整个码头观赏有了更好的视野，对客人来说也是个很好的住店体验。

The Met Hotel is a 5-star hotel entirely designed by Zege Architects for the Chandris Group.It has become a new urban landmark attracting people to an industrial environment that was hardly ever visited by the wider public before.It contributes to the development of the port and sets an example of how such areas can be given new life. The building is in a dialogue with its context:the sea and the docks lend it its special form.The wavy surfaces of its facades —made of black and white marble— challenge the standard hotel typology and the zigzag lines of the bar,the reception desk and the restaurant give a sense of movement that prevails over the straightness of the enveloping spaces.Big windows and transparent bathroom walls in the rooms make the view of the port an integral part of the design and the visitors' experience.

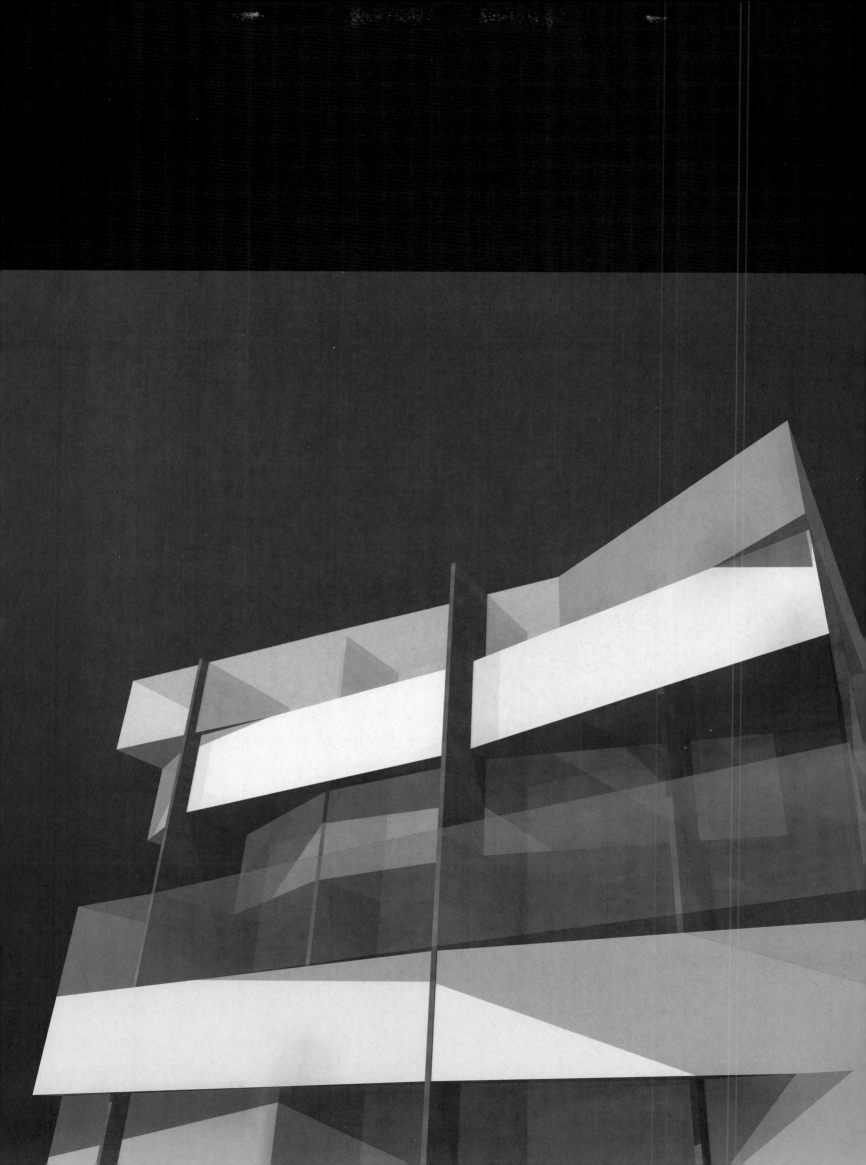

项目大胆应用黑色作为空间的主色调，将黑白对比的空间设计技巧发挥到极致。所有环境设计的要素，围绕色彩的对比而展开。材料与构造、装修与设施、陈设与配饰的选择与布置精准到位。

——郑曙旸
（国务院学位委员会设计学学科评议组组长、清华大学美术学院教授、博士生导师）

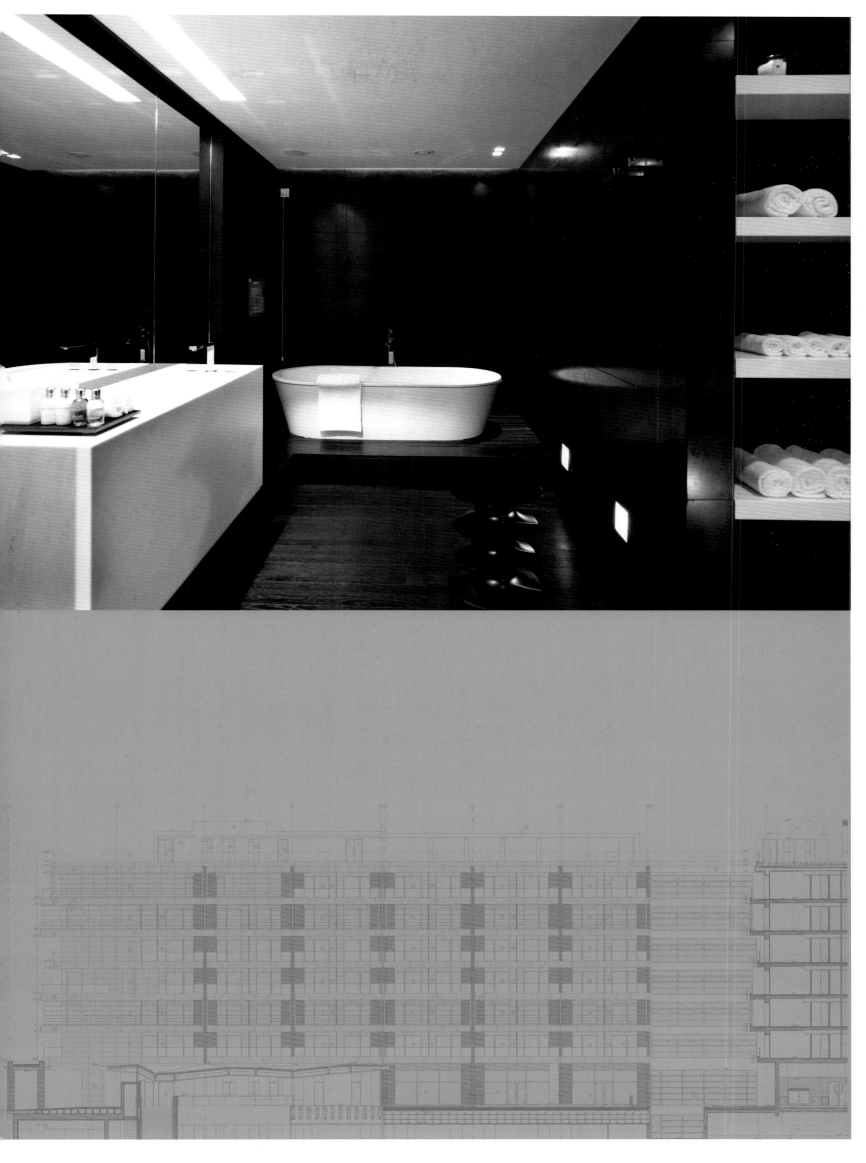

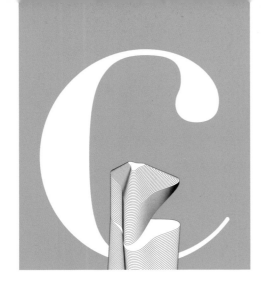

提名奖

The Nomination for Best Design Award of Hotel

最佳酒店设计

Sanya Haitang Bay No.9 Resort Hotel
三亚海棠湾9号度假酒店

设计者 | Designer

YANG BANG SHENG Hotel Design Consulting Ltd.
杨邦胜酒店设计顾问公司

设计说明 | Design Illustration

三亚海棠湾9号度假酒店地处美丽的国家海岸海棠湾片区，与蜈支洲岛隔海相望，总建筑面积8万平方米，是中外游客休闲度假胜地。酒店整体以绿色作为主色调贯穿设计之中，运用简洁明快的设计手法将现代中式风格与热带海岛风情完美结合。并巧妙融入海南黎族文化，以国际化的品质突显独特的"泛东方"度假酒店魅力。

设计师还从海南独特的黎族文化中汲取灵感，将黎族农村劳作的工具——打谷桶演变成大堂总台的背景，与鸟笼样式的巨型装饰灯相呼应。闻名于世的黎锦纹样被提取、凝练成富有现代感的装饰图案，巧妙地运用在地面、墙面、屏风、家具、地毯之上，体现出浓郁的地域文化气息。特别值得一提的是，设计中倡导环保节能理念，公共区域内，开敞的折叠门取代了生硬的外墙，将自然的阳光、空气引入室内，使室外的园林、海景与室内空间融为一体，营造出自然、舒适、浪漫的度假空间。

Sanya Haitang Bay No.9 Resort Hotel, which is located in beautiful national Seashore Haitang Bay Area, faces Wuzhizhou Island across the sea. The total construction area is 80,000 square meters. It is a leisure resort for domestic and foreign tourists. Green is the main hue, the designer make well use of simple and neat design techniques combining with modern Chinese style and tropical island characteristic. Hainan Li culture is cleverly integrated into the hotel design, and through the unique international quality to highlight "Pan Orient" resort charm.

The designer is inspirited by the Hainan Li unique culture, taking the Li rural labor tools—threshing bucket as the background of the lobby, which echoes with the decorative lights, a giant bird cage style. The famous Li Jin patterns are extracted, and then condensed into very contemporary decorative patterns. The designer makes clever use of these decorative patterns and uses them on floors, walls, screens, furniture and carpets, reflecting the strong regional culture. It is particularly worth mentioning that the green energy design concept, such as in the public areas, the open folding doors replace the rigid walls, making the natural sunlight and air shine into the room, the interior space blend into the outdoor garden and sea, creating a natural, comfortable and romantic holiday space.

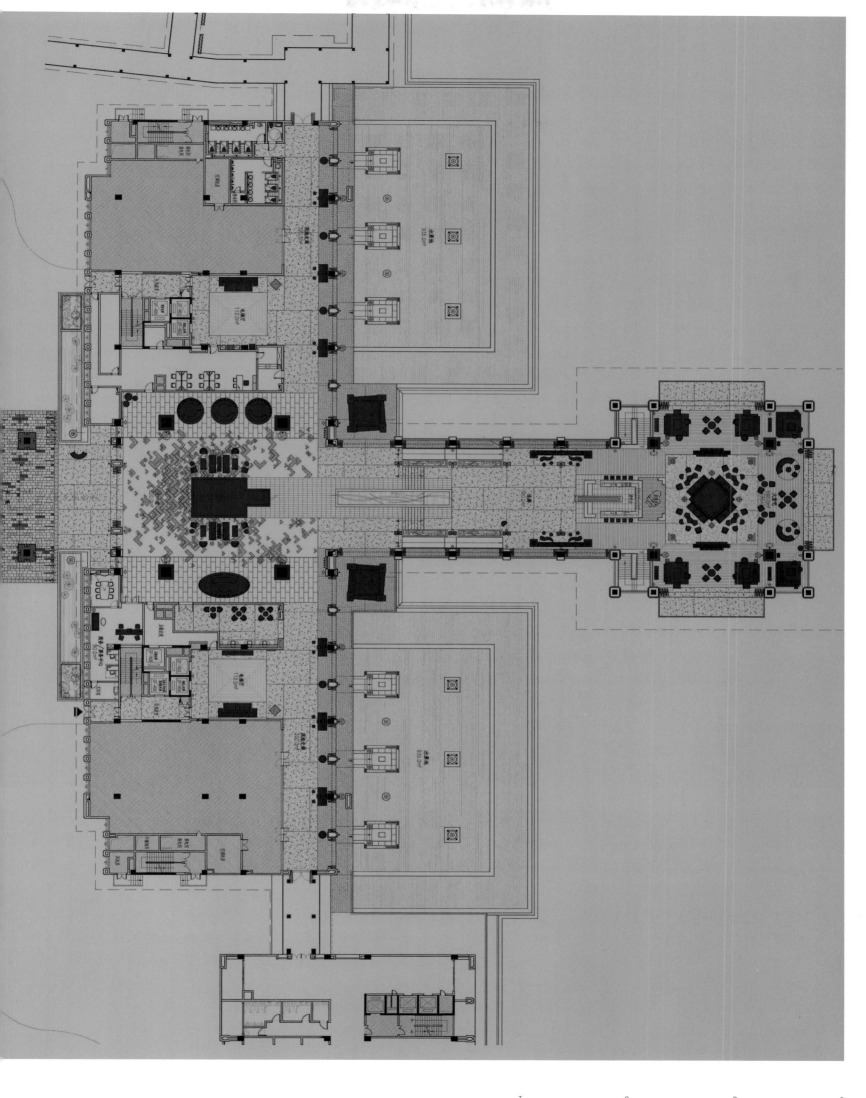

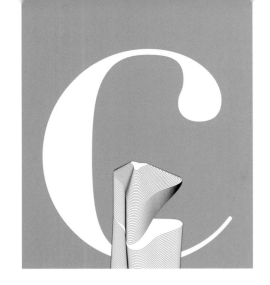

提名奖
最佳酒店设计
The Nomination for Best Design Award of Hotel

Shanxi Zha Shui Luyuan International Hotel
陕西柞水麓苑国际大酒店

设计者 | Designer

Shenzhen Concentric League Decorative Design co., Ltd.
深圳市同心同盟装饰设计有限公司

设计说明 | Design Illustration

酒店设备齐全，装修质朴典雅，集东南亚建筑风格和现代设计风格为一体，无不透着久违的悠闲与自由。酒店A区主楼共八层，拥有不同风格类型客房近260间（套），温馨舒适，别具一格；区内设有大堂吧、全日制西餐厅、行政酒廊和商务中心，您可在细品香浓咖啡、体验异国美食的同时，饱览山中美景，聆听潺潺河水之音。B区设有中餐厅及VIP包房、日韩料理、宴会厅、多功能会议室、室内恒温游泳池、康体中心、SPA会所等服务设施，灵活便捷的会议场地，强大的服务设施功能，定会让您的工作或度假别有体会和倍感轻松。

The hotel offer comprehensive facilities.It combines Southeast Asia-style with Modern-style and shows comfortable and free to us.Zone A total of eight-storey main building of the hotel,with standard rooms,executive rooms,executive suites and presidential suites,260 rooms of different types of styles (sets) and the hotel lobby,full-time Western restaurant,executive lounge and business center.Rooms of the hotel are appointed to provide ultimate in comfort.You can drink coffee,experience exotic food,enjoy hill-view or listen to voice of the river at the same time.Zone B has a Chinese restaurant and VIP rooms,Japanese&Korea restaurant, banquet hall, multi-functional meeting rooms, indoor heated swimming pool,sports center,SPA club and other service facilities.Flexible meeting space and strong service function will make your work or holiday relaxed.

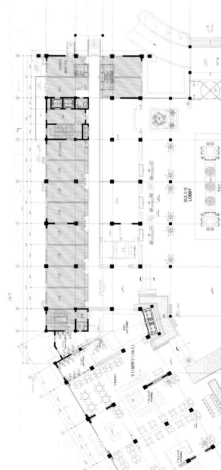

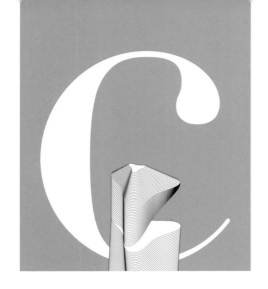

提名奖

最佳酒店设计
The Nomination for Best Design Award of Hotel

Dali N° Hotel
大理 N° 酒店

设计者 | Designer

纳杰 Na Jie

设计说明 | Design Illustration

大理 N° 酒店是一个集聚创意的场所，充满了很多优秀的设计理念。N° 酒店与大理其他酒店不同，不是依山傍水，并没有拥有大理得天独厚的自然风景。所以我们不得不开创另一个风景——设计。N 的来源也是一个创意的开始。N 除了代表投资方外，更代表了我们数学中常常说到的 ∞（无穷）。N° 酒店对于旅客来说，没有引人入胜的风景，所以我们必须创造一个能让旅客停留的条件。一成不变的房间没有视觉感受的挑战，而一个独特的风格可以让人们耳目一新。即便是经常入住 N° 酒店，相信每次都会有不同的感受。

Dali N° hotel is a gathering place for creative,full of a lot of good design. N° hotel is different from other hotel in Dali,not a mountain,did not have the unique natural scenery of Dali.So we have to start another scenery—design. N° the style of the hotel location is,of course,the key factors.Both decided to do a design hotel,will put the design to do the most perfect. N° hotel for passengers,no spectacular scenery,so we should create a condition allows passengers to stay.No visual perception challenge remains the same room,and a unique style can make people find everything new and fresh.Even if is often stay in hotels N°,believe that every time I have different feelings.

最佳**餐饮空间**设计奖
BEST DESIGN AWARD
OF DINING SPACE

最佳餐饮空间设计艾特奖 Best Design Idea–Tops Award of Dining Space
吕靖

最佳餐饮空间设计提名奖 The Nomination for Best Design Award of Dining Space
Minas Kosmidis / 俞怀德 / 毛赟 / 邹巍、高波

艾特奖

Best Design Idea-Tops Award of Dining Space

最佳餐饮空间设计

Hangzhou NIMO Western Food&Bar
杭州尼莫 NIMO 西餐酒吧

设计者 | Designer

吕靖 Lv Jing

设计说明 | Design Illustration

整个设计主题由海洋演变而来，采用灰色为主题色，蓝色适当点缀。由于主体空间偏狭长且层高较高，设计从中轴线将狭长的空间切割成对称的两个体块。餐厅由酒水吧台为正中心作为一个完整的主体，酒水台正对小空间聚会区，分割了整个空间。由于空间狭长且偏高，本案采用了两个巨型台灯及两侧树的延伸造型来拉近层高，使得空间饱满且不易一眼望穿。

The whole design theme is evolved from the ocean element, gray is the themed color tone, and blue is interspersed. Because the main space spreads very long and narrow, and the storey is high, the designer cut the narrow and long space into two symmetry pieces by the axis. Restaurant Wine Bar is the main part of the while space, the wine bar faces small space party area, carving up the entire space. Due to narrow and high space, the designer takes two giant lamps and the extended shape of trees on both sides to narrow the storey, making space full and it's difficult to wear out for customer's eyes by gazing.

项目准确地界定了西餐酒吧所需的设计概念发展方向。运用深色调空间背景与极具特点的装饰构架,共同营造出静谧兼具诙谐的空间氛围。所选家具与陈设物同样符合场所需求。

——郑曙旸
(国务院学位委员会设计学学科评议组组长、清华大学美术学院教授、博士生导师)

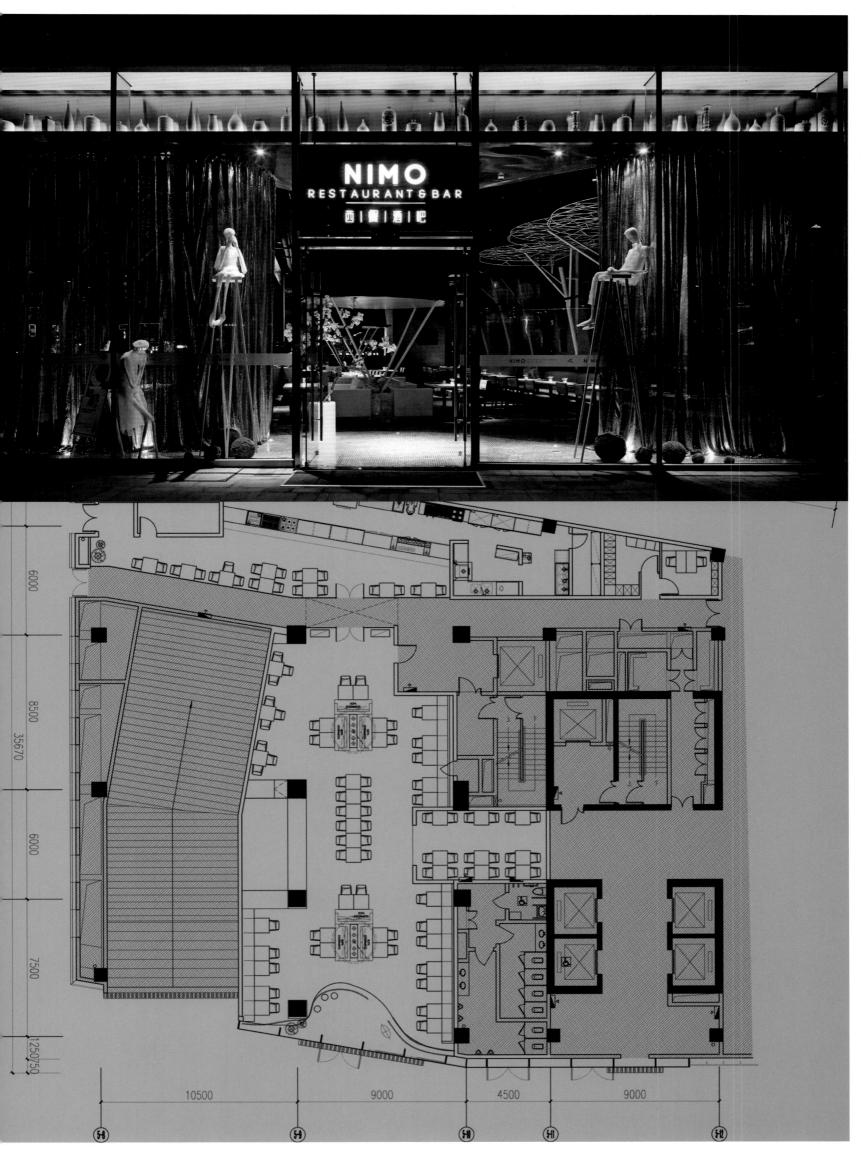

提名奖

最佳餐饮空间设计

The Nomination for Best Design Award of Dining Space

O-13

设计者 | Designer

Minas Kosmidis

设计说明 | Design Illustration

设计师采用了"海洋"元素作为设计主题。以"水"为基本链接元素，让设计元素与每一季艺术时尚潮流元素进行互动。几何形状的设计，与菱形壁纸和饰品所采用的颜色相呼应，蓝色的大海，蓝色的天空，红色的高黏合度的核桃木质表面，这些核桃木的设计即是构造元素的体现，也具有强烈的象征意义，另外设计师还采取鱼鳞形状的设计来达到想要的效果。

Communicating vessels with the "water" being the basic element to this interaction; like the trends—the waves of art of the fashion of the design each season. Geometric shapes, with a strong presence of the rhombus both on wall-papers and fabric with the colors of passion, the blue sea, sky, and red intensity combined with wooden surfaces from walnut sometimes like constructed elements and sometimes taking shapes of intense symbolism like fish scale that consist elements—parts of the puzzle of this area.

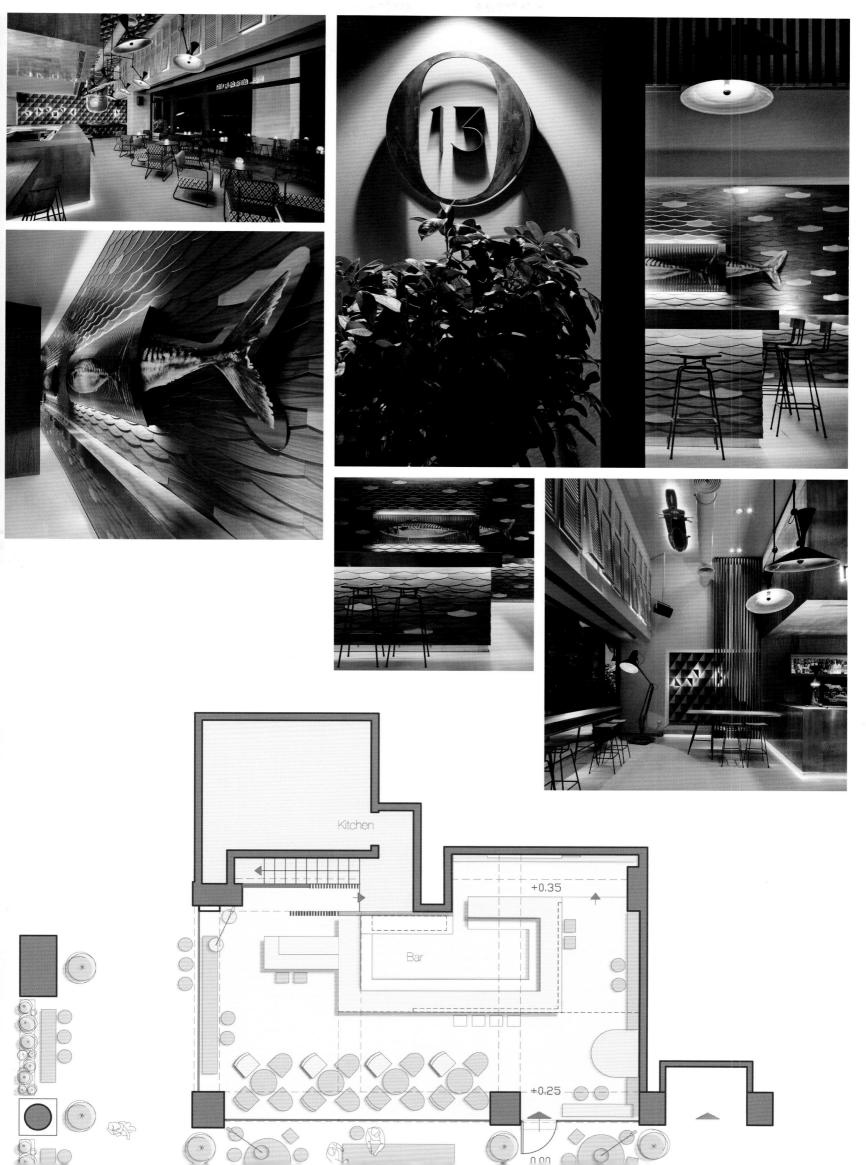

提名奖
最佳餐饮空间设计
The Nomination for Best Design Award of Dining Space

Mengjiang Noodle Restaurant
猛将面馆

设计者 | Designer

俞怀德 Yu Huaide

设计说明 | Design Illustration

空间运用大量的回收旧木，加工出各种木方、板材，特意保留了手工痕迹，进入空间让人们体验到自然气息。在这里，室内空间成了自然的媒介，人们通过建筑空间能够感知自然的气息。熟悉的材质、简单的色系，多了些许亲切。

The designer makes full use of the recycling old wood, and processes a variety of wood, sheet metal, specially retains the manual trace. When you step into the space, you can experience the natural flavor. In this space, the designer takes the interior space as the medium, people can perceive natural atmosphere through the architectural space. The regular materials, simple colors create a cordial space.

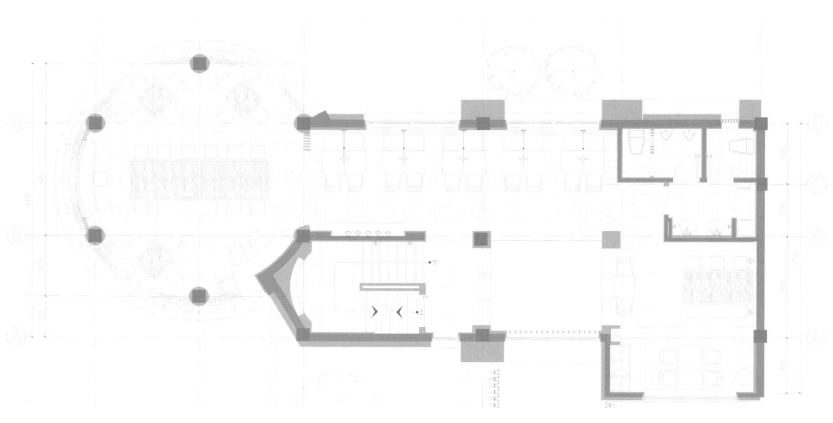

提名奖

最佳餐饮空间设计
The Nomination for Best Design Award of Dining Space

Kaogu BBQ
烤古烧烤

设计者 | Designer

毛赟 Mao Yun

设计说明 | Design Illustration

本案所呼唤的是一种对原始人类行为的回归。一楼的座位以条形序列的排放形成视觉上的纯粹之感。二楼摆了一张大桌，其他位置以L形吧台的形式出现，使得人在空间里更加随性，更加便于互动交流。烤古品牌的第一个"烤古"出发现场为人类的起源河姆渡遗址。我们选用了遗迹中出现的木料、稻壳、陶土罐，根据地质层的分层，体现在门店的墙上，同时，用了考古发掘常用的施工照明灯组成波浪形状的装饰组灯。

The project focuses on the original construction of the most fundamental of human behavior regression, the seats in the first floor range in an banding manner and form a sense of visual; in the second floor, there is a large table, a L-shaped bar, making people much more casual, and it's easier for people interacting with each other in such a space. The designer hopes to pursue spiritual properties of the material, the brand dates from the human origins Hemudu site, the designer selects wood columns appeared in ruins style architecture, including the rice husk, clay pot reflected on the wall give the stores under stratified geological layers, meanwhile the undulating set of lights is consist of the lighting commonly used in the archaeological excavations.

提名奖

最佳餐饮空间设计

The Nomination for Best Design Award of Dining Space

Nongzili Restaurant
弄子里餐厅

设计者 | Designer

邹巍、高波 Zou Wei, Gao Bo

设计说明 | Design Illustration

设计以需求出发，以一种跨界的思维和中西文化混搭的手法，试图让设计能够与环境交流。尽量去接近自然，白色的砖墙、干枯的树枝、栖息在屋檐上的鹦鹉、小鸟、景德镇的陶瓷、意大利风格的个性定制地砖、欧洲街头的油画，无处不透着对自然和艺术的向往。在陈设运用上，强调了对立与和谐，突出空间表情的丰富性。通过装饰元素及中西文化之间存在的差异碰撞出空间的激情，让食客们产生对空间的眷恋。

Considering the design needs, the designer make the design communicate with the environment by using the cross-border approach and mixing cultures. In order to get close to the nature, the designer use the white brick walls, dry branches, the parrot perched on the roof, birds, Jingdezhen ceramics, Italian-style personalized custom tiles, Europe painting to make the ubiquitous nature and art longing. In the term of the furnishings, the designers focus on opposition and harmony, to highlight the richness of space expression. The space provides the diners nostalgic mood, the passion of space is expressed by decorative elements and collision between Chinese and Western cultures.

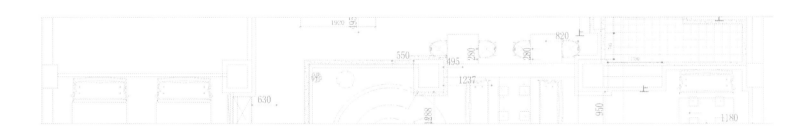

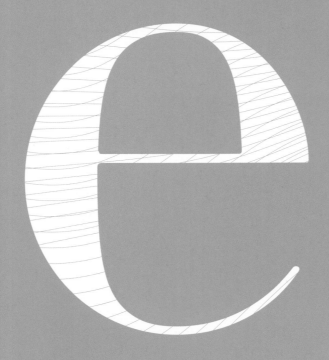

最佳娱乐空间设计奖
BEST DESIGN AWARD
OF ENTERTAINMENT SPACE

最佳娱乐空间设计艾特奖 Best Design Idea–Tops Award of Entertainment Space

王践（宁波矩阵酒店设计有限公司）

最佳娱乐空间设计提名奖 The Nomination for Best Design Award of Entertainment Space

多维设计事务所 / 大展设计顾问有限公司 / 张礼斌 / 福州维野餐饮娱乐空间策划设计机构

艾特奖

最佳娱乐空间设计
Best Design Idea-Tops
Award of Entertainment Space

Mansion Entertainment Palace
大公馆娱乐殿堂

设计者 | Designer

宁波矩阵酒店设计有限公司
王践 Wang Jian

设计说明 | Design Illustration

大量运用易加工成型、可再生且达到防火等级的铝材，工业化的流程大大降低生产安装成本。大量运用幻彩及陶瓷马赛克这一古老而神秘的装饰建材，利用马赛克多变的色彩和细腻的质感装点空间，超大幅面的图案无须过多装饰。一幅画面就是一个故事，一件艺术品。让传统的陶瓷、玻璃类建材与新型的金属类材料在同一空间中和谐共生。空间体量宏大，尺寸与维度都十分震撼又不乏舒适。色调与装饰风格自成一派，与传统娱乐空间形成鲜明对比，激发了消费者强烈的好奇心。简约整体的装饰手法和坚固耐用的装饰建材也大大降低了经营者的维护、保养成本。

Easily processing and forming, renewable and reaching the fire rating of aluminum materials were been used a lot .Industrialization process greatly reducing production cost.Iiridescence and ceramic Mosaic,those ancient and mysterious decorative building materials,were used in great quantities.Take advantage of Mosaic changeful color and delicate texture to deck space,no need too much decoration of large sized pattern.One picture tells a story,one picture is a work of art.Let the traditional ceramic,glass building materials and new type of metal materials harmoniously coexist in the same space. Huge spatial volume,size and dimension are stunning and comfortable.Color and the decoration style is unique in its own style,in sharp contrast with the traditional entertainment space,inspiring consumers' strongly curiosity. Whole simple adornment gimmick and durable decorative materials can also greatly reduce the operator's maintenance and maintenance' costs.

娱乐空间设计成功的关键在于确立消费对象的文化定位。该项目精准地把握消费心理、消费习惯，并针对娱乐空间特殊的服务流程和需求，选取相应的环境设计要素，营造出符合娱乐潮流趋势，追求高雅文化意趣的空间场所。

——郑曙旸
（国务院学位委员会设计学学科评议组组长、清华大学美术学院教授、博士生导师）

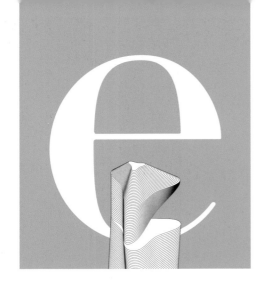

提名奖

最佳娱乐空间设计

The Nomination for Best Design Award of Entertainment Space

Chengdu Magic Volume Sales KTV
成都奇幻量贩 KTV

设计者 | Designer

DODOV Design Office
多维设计事务所

设计说明 | Design Illustration

本案设计采用梦境、奇幻为主题，是当代年轻人追崇的时尚元素，也符合年轻消费者的精神和心理诉求。设计上摒弃了量贩式单一直白的色调，将金、银、酒红、深咖、暗紫、深蓝等多种颜色巧妙融合，令空间色彩丰富，层次突出，具有视觉冲击感。同时该KTV把平面设计作为空间设计的一大主题和亮点，采用一个包间一个主题画的形式，打造奇幻量贩式KTV。

The theme of this design case is dream and fantasy,this is the fashion element that contemporary young people chasing after,it also accord with young consumers' spirit and psychological demands.Design abandoned the volume type single straight white tonal,combines the gold,silver,wine red,deep coffee,dark purple,dark blue with other colors skillfully, makes spatial color abounded,creates a outstanding level and impacts the sense of vision.The graphic design been a theme and bright spot of the KTV spatial design at the same time.In the form of one room one theme painting,creating the Magic Volume Sales KTV.

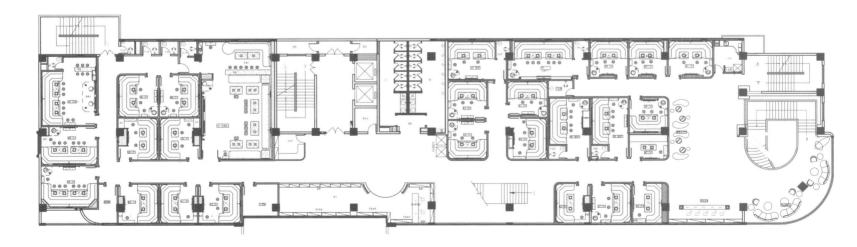

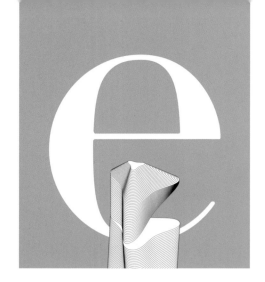

提名奖

最佳娱乐空间设计

The Nomination for Best Design Award of Entertainment Space

Shenyang Jade Collection Club
沈阳金玉荟会所

设计者 | Designer

Dazhan Design Consultant Co., LTD
大展设计顾问有限公司

设计说明 | Design Illustration

空间整体由红酒酒吧兼设小型餐饮两部分组成，小型的餐饮形式活跃补充了空间的多样性，相对于餐饮设计的平稳中见精致混搭，红酒吧设计要张扬多了。本身公共区具有先天的高架空间，这给了设计师无穷的想象空间，设计风格比较模糊，较多地运用了重复的格子造型，从有序的阵列到无序的组合。移动的黑漆门更增加了空间的光影变化，塑造了不同的空间感受，精致的木饰面造型与粗糙的青石板、原木柱、黑镜钢等材料有机结合起来。在墙面上描有大面积的艺术油画，使整个空间在具有设计感的同时增加了更多的人文情怀。

Whole space consists of two parts,wine bar and a small restaurant,small catering form actively riches space diversity,and relative to the smooth and delicate mix built catering design,red bars clubs design is much more publicized.Public area have innate lifts rack space itself,which gives the designer infinite imaginary space,this part of the design style is blurred,many repeating in the grid model,from the orderly array to the disorder.Mobile black paint door increased the light of the space,shape different space feeling,fine wood veneer modelling and rough,green flag,the original wooden poles,black mirror steel materials integrate an organic union,a large area of the art painting drawn on the wall,let the whole space be filled with design feeling while be added with more humanistic feelings.

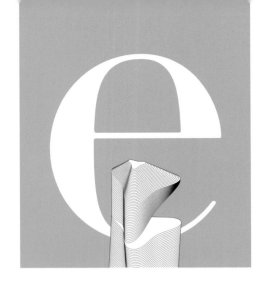

提名奖

最佳娱乐空间设计
The Nomination for Best Design Award of Entertainment Space

Jinshan Mansion KTV
金山公馆 KTV

设计者 | Designer

张礼斌 Zhang Libin

设计说明 | Design Illustration

大的结构和思路来源于星舰，但高于星舰造型本身。于是把星空的理念和飞船的意境提炼出来，故整体里面造型多无规则带动感流线型，这样看上去无规律，其实都有无数的计算暗含其中，隐喻星空、猎奇、探知、叛逆，特别是最后一点——叛逆，这才是年轻和时尚的本质表现。

The idea of structure and conception comes from the starship,greater than the starship,than the designer wants to derive the concept of the stars and the artistic conception of the shuttle itself,so the whole inside have no sense of rules to drive more streamlined shape,it looks random rate,it has implied the countless calculation. Metaphor the starry sky,novelty,detection,rebellious,especially the last point,rebellious,this is the essence of the youth and fashion.

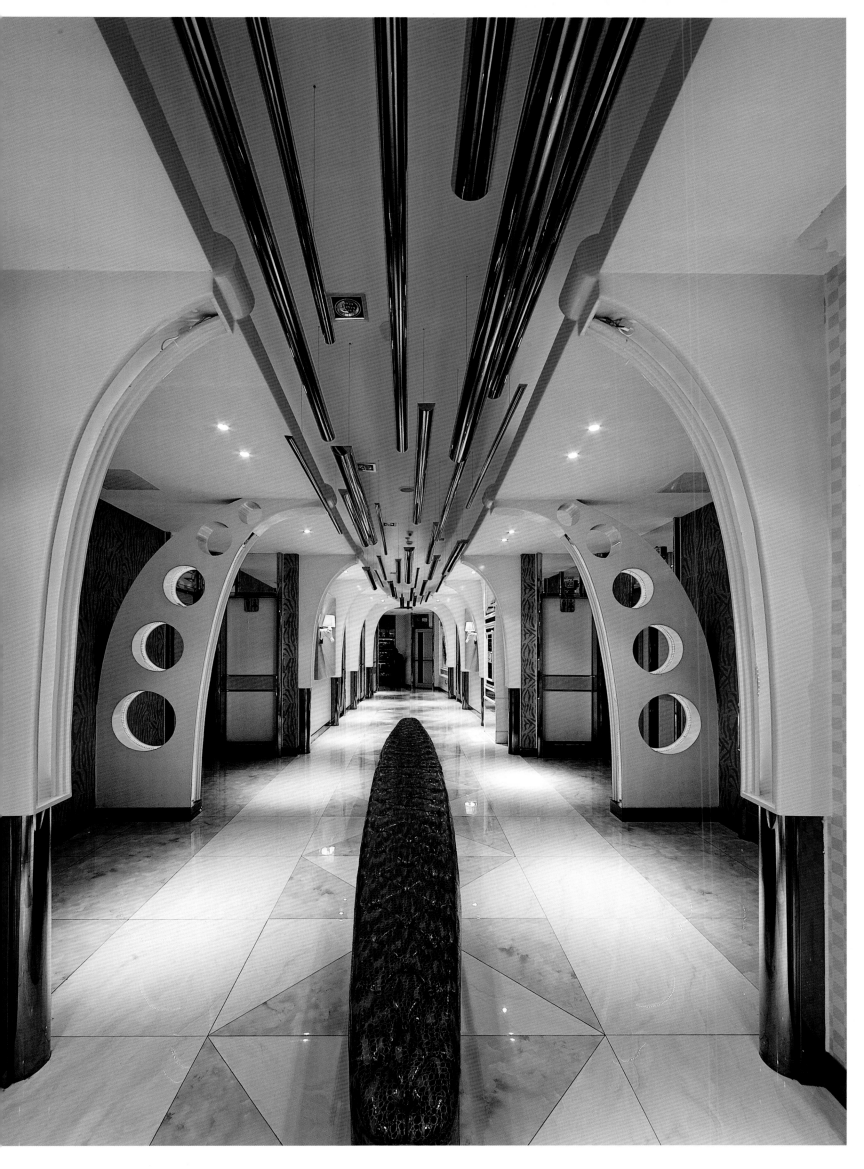

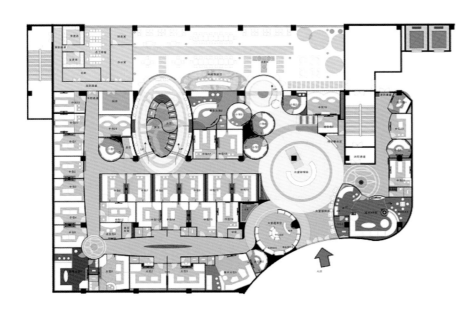

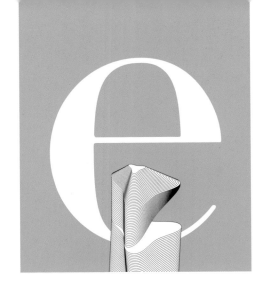

提名奖

最佳娱乐空间设计

The Nomination for Best Design Award of Entertainment Space

Jiangxi Yongxiu Oriental Masai Music Club
江西永修东方玛赛音乐会所

设计者 | Designer

Fuzhou Weiye dining entertainment spatial planning design agency
福州维野餐饮娱乐空间策划设计机构

设计说明 | Design Illustration

借助东方情怀的建筑，以浮夸的装饰去凸显这些元素，赋予这些装饰效果新的特质，呈现出完全不同于过往与未来的新面貌。当外观手绘方案出来感觉还不错，于是贯穿于室内将这种手法进行下去。

Taking advantage of the building with oriental feelings, with grandiose adornment to highlight those elements, giving these adornment effect new characteristics, present a completely new look different from the past and future. When hand-drawn scheme came out fine, this technique would continue to use in interior work.

最佳**展示空间**设计奖
BEST DESIGN AWARD
OF EXHIBITION SPACE

最佳展示空间设计艾特奖 Best Design Idea—Tops Award of Exhibition Space
HENN Architect

最佳展示空间设计提名奖 The Nomination for Best Design Award of Exhibition Space
Un Studio / 彭征 / LIKEarchitects / 艾恩迪（北京）室内设计有限公司

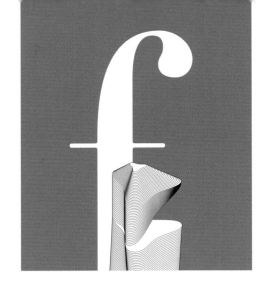

艾特奖

Best Design Idea-Tops
Award of Exhibition Space

最佳展示空间设计

Porsche Pavilion at the Autostadt in Wolfsburg

设计者 | Designer

HENN Architect

设计说明 | Design Illustration

该建筑是独一无二的，它的建设卓越不凡。此馆还有一个历史的象征意义，从一开始保时捷和大众的联系，并将在未来继续不可分割的关系。保时捷公司首席执行官马蒂亚斯穆勒这样说："作为大众，全球领先的汽车终端和沟通平台，我们会为顾客深入了解我们的品牌、价值和理念。"在保时捷展区，我们开创了Autostadt的历史新篇章，增添奥托楼Wachs，Autostadt的领导者。

The building is unique and its construction is extraordinary. This pavilion also has a symbolic and historical dimension, as it hints at the common roots through which Porsche and Volkswagen have been connected from the very beginning and will continue to be connected also in future, says Matthias Müller, CEO of Porsche AG. "As a worldwide leading automobile destination and communication platform for Volkswagen, we provide insights into our brands, values and philosophy for our guests. With the Porsche Pavilion we start a new chapter in the history of the Autostadt", adds Otto F.Wachs, Director of the Autostadt.

Forms translating Porsche's legendary aerodynamico. Enhancing expression or volumes and changing light in the landscape. 从该项目中，设计语言的表达，卷轴的体现以及景观灯光的变化都是一大亮点。

——Ilhan Zeybekoglu 依鲁翰·泽碧克格鲁
（美国 ZNA 泽碧克建筑设计事务所创始人、美国哈佛大学建筑学院教授）

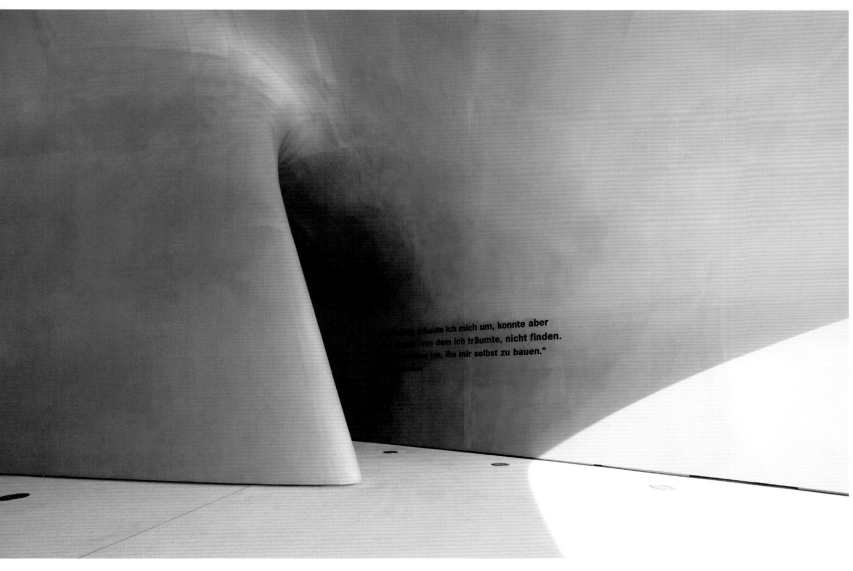

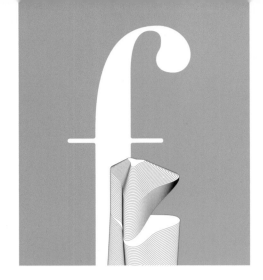

提名奖

最佳展示空间设计

The Nomination for Best Design Award of Exhibition Space

Motion Matters

设计者 | Designer

Un Studio

设计说明 | Design Illustration

主要聚焦于重点项目之外，借由五个主题从而生成了更详细的视图，五个主题通过展览空间沿三维丝带将项目结合在一起 。这些主题由无数个很微小的，来自展出项目之间包含更多内含的灵感及其多种关系的图片构成。这是辩论和实现过程，并提供了解 UN Studio 实践知识驱动的性质。

In addition to the primary focus on these key projects,a more detailed view is generated by means of five thematic threads which bind the projects together along a three-dimensional ribbon which meanders through the exhibition space.These threads consist of numerous small images which afford the reading of the various relationships between the exhibited projects within a larger context of inspiration,debate and realisation processes and provide insight into the knowledge driven nature of UN Studio's practice.

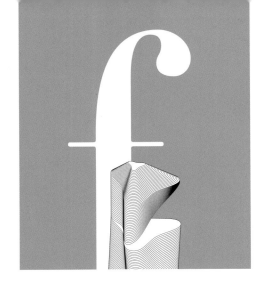

提名奖
最佳展示空间设计
The Nomination for Best Design Award of Exhibition Space

Metamorphosis—Victory Garden Sales Center

变形计——凯旋荟销售中心

设计者 | Designer

彭征 Peng Zheng

设计说明 | Design Illustration

售楼部的设计巧妙地应用了山形折面，以"石"对应山，突出场地优越的地理环境。室内空间将原本单一的空间分解与重组，其折叠形式与质感呼应建筑外形特征，带来连续的空间体验。室内的家具满足功能需求的同时，如雕塑般与空间特质相得益彰。户外景观广场作为售楼部功能的延伸，将人们的视线带回场地独特的地理环境。

The design of sales department takes advantage of the fold surface shape skillfully, highlight the ground of superior geographical environment. Indoor space decompositioned and restructured originally single space. The folding forms and textures echo building shape characteristics, bring continuous spatial experience. Indoor furniture meets the functional requirements, as sculpture and spatial characteristics reflected akira. As an extension of the sales department, outdoor landscape square bring people's sight line back to the unique geographical environment.

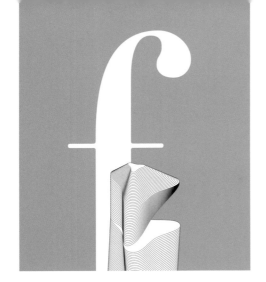

提名奖 | The Nomination for Best Design Award of Exhibition Space
最佳展示空间设计

The Andy Warhol Temporary Museum

设计者 | Designer

LIKEarchitects

设计说明 | Design Illustration

室内被设计成一个传统封闭的空间，完全被连续的墙壁所定义，被塑料屏幕的透明盖子所保护，允许外部光线进入，并保证博物馆和购物中心这两个对视空间的透视关系。

The interior was designed as an enclosed introspective space, entirely defined by continuous walls, benefiting from a transparent cover in plastic screen, allowing light to enter from the exterior and assuring the visual relationship between the two confronting spaces—museum and shopping mall.

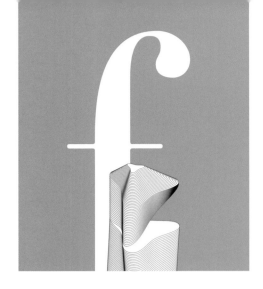

提名奖

最佳展示空间设计

The Nomination for Best Design Award of Exhibition Space

Triple Royal Sales Center
三合御都售楼处

设计者 | Designer

IND(Beijing) Decoration Design Co.,Ltd
艾恩迪（北京）室内设计有限公司

设计说明 | Design Illustration

结合客户群与地块的独特地缘，结合地域、人文、审美、项目定位等多个要素来考虑。可以看到售楼处的平面通过合理的规划让空间变得有趣、轻松、互动，给客户更多的体验和享受，赋予空间更多的文化内涵。因为每个楼盘的销售行为是时刻性的，从延续的时段性来说，它应当是一个社交圈。当客户进入这个圈子，找到了共同点，从一至无穷的概念，这就是我们所说的"圈层营销"的本质。

Taking consider on customers and unique geographical plot, and other factors such as region, humanity, aesthetic appreciation, project orientation. Sales Office flat makes space funny, easy, interactive by reasonable planning, gives customers a wider range of experiences and enjoyment, gives space more cultural connotation. Since every real estate sales activity happens at a time, for extended periods. It should be a social circle, when a customer enters the Club and finding common ground, from the concepts of reaching one to infinity. This is what we call a "circle marketing" of human nature.

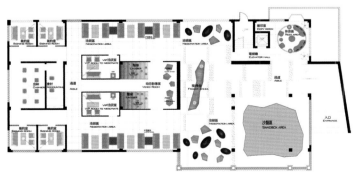

最佳会所设计艾特奖 Best Design Idea–Tops Award of Club

何武贤

最佳会所设计提名奖 The Nomination for Best Design Award of Club

深圳市新冶组设计顾问有限公司 / 深圳市阮斌设计有限公司 / 成志、胡俊峰 / 徐明

艾特奖

最佳会所设计
Best Design Idea-Tops Award of Club

CrossRoad | The Turning Point
界 | 转折 人文会所

设计者 | Designer

何武贤 He Wuxian

设计说明 | Design Illustration

本案企图整合传统式街道型的店家模式，透过系统化且多元性的空间组合，开创台湾第一家复合式殡葬礼仪会所空间。"界|转折"是一个具有生命哲学观的主题会所。本案在发想阶段，收到学生从日本寄来的明信片，随手拈来，折出了空间的界面，空间转折光影界分，人生转折因缘微妙。人生最终的转折犹如空间的转折，此没彼生，彼没此生，片纸转折思念祈福，生死界分人间净土。

Using a systematic yet versatile space design, this project integrates traditional retail vendors into the first multi-function funeral service hall in Taiwan. "CrossRoad: The Turning Point" is a reception hall inspired by a this profound life philosophy. When this project was conceived, I received a postcard sent from my student in Japan. As I played with the postcard, bending, folding, creasing, I realized I was creating a space. The light casted shadows within this space, creating borders, flections, and intersections, much like the bends and turns of life. Death in this world simply means birth in another. Through the simple act of paper-folding, we learn that the loved ones leave their mortal bodies only to embark on a new journey.

An inviting and engaging expression of the leisure time.Delicate use of colors and materials. 休闲时间邀请亲朋相聚的绝佳场所,此作品在色彩与材料应用上独具匠心。

——Ilhan Zeybekoglu 依鲁翰·泽碧克格鲁
(美国 ZNA 泽碧克建筑设计事务所创始人、美国哈佛大学建筑学院教授)

提名奖
最佳会所设计
The Nomination for Best Design Award of Club

Shenzhen Belin Facial Club
深圳贝琳美容院

设计者 | Designer

Newera Design
深圳市新冶组设计顾问有限公司

设计说明 | Design Illustration

不管是前台接待区，还是中庭的休息区都采用了不规则的几何空间形态。从墙板到隔断、吊顶、吊灯……步入其间，仿佛走进了外星人的飞船，在形态的不规则之中，却随处可见仿真花草的点缀，一切的不规则，却因为这份绿意，变得浪漫起来。解构之于生活，是对习惯的分解与重组；解构之于设计，是打破规则，重建空间秩序的思考。而解构之于生活中的绿洲，则代表着紧张忙碌面孔之外的另一种生活主张，在都市生活中寻找美丽的人，更能明白设计师这层意味深长的慰藉吧。

Both in the reception area and the atrium lounge area,the designer adopts a form of irregular geometric space,including the wall,the partition wall,the ceiling and chandelier…when you step into this space,it seems that you are in spaceship of the alien,in this irregular space,you have simulation flowers dotted everywhere, all the irregularities,with these flowers and grasses, becomes an romantic space.Deconstruction to life, it is customary decomposition and reorganization;deconstruction to design it is breaking the rules and the reconstruction of re-thinking spatial order;deconstruction to the relax life,which represents another hectic life,looking beautiful people in urban life,having a better meaningful understanding for designer.

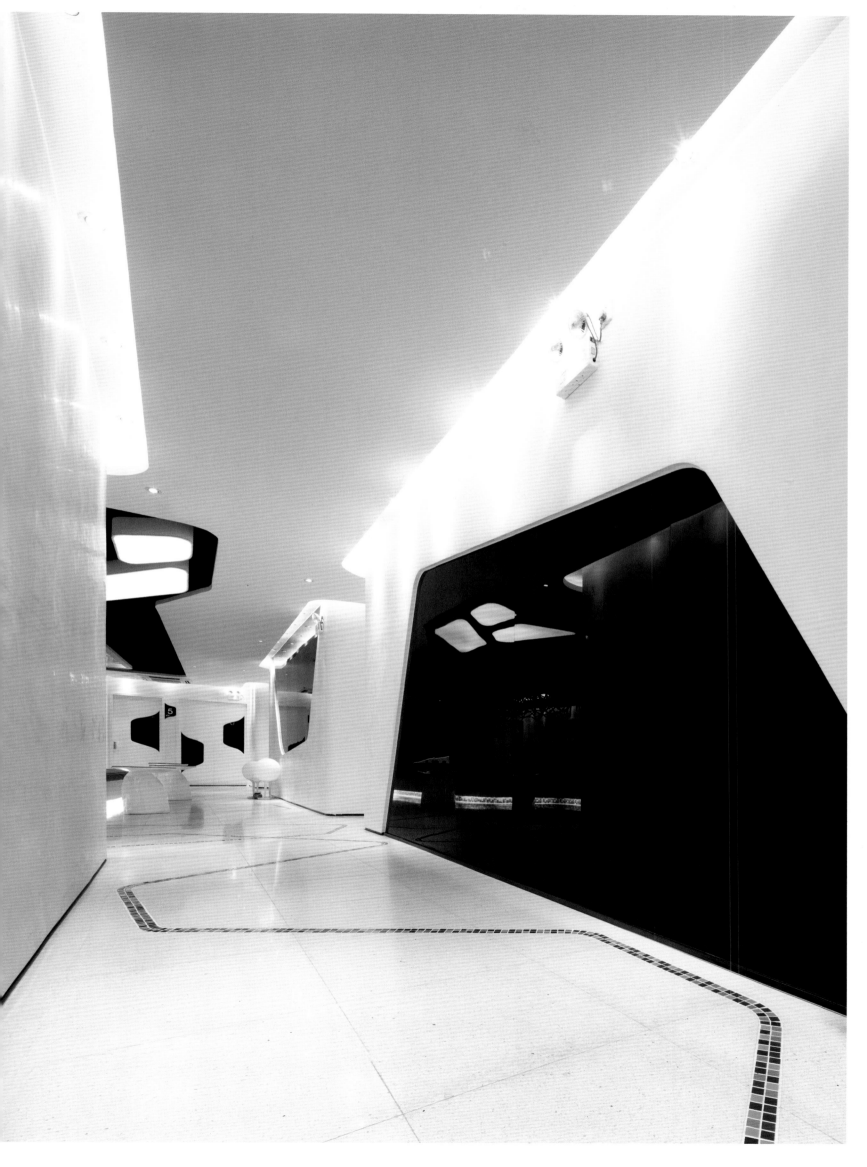

提名奖

最佳会所设计
The Nomination for Best Design Award of Club

Shenzhen Galaxy Danti Taiji Club
深圳星河丹堤太极会馆

设计者 | Designer

Shenzhen Ruan Bin Design Co., Ltd
深圳市阮斌设计有限公司

设计说明 | Design Illustration

太极拳含蓄内敛、连绵不断、以柔克刚、急缓相间、行云流水的拳术风格使习练者的意、气、形、神逐渐趋于圆融一体的至高境界，而其对于武德修养的要求也使得习练者在增强体质的同时提高自身素养，提升人与自然、人与社会的融洽与和谐。对于这一主题的表现我们摒弃了过于直白的太极拳雕塑和图像资料，寻求符合太极理念意象的饰品，使观者有更深层次的触动和感受。

The characteristic of Tai Chi is implicit,quietness,continuous,conquering the unyielding with the yielding,acute and slow & smooth.The exerciser of the Tai Chi makes the harmony of the body and spirit as well as its requirements for morality cultivation practices and physical fitness to enhance the harmony of man and nature,man and society. For this topic,the designer rejects the Tai Chi sculpture and pictorial materials,seeking to meet the image of Tai Chi philosophy of jewelry,making the viewer deeper touch and feel.

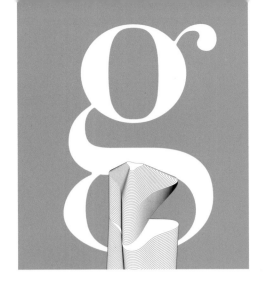

提名奖

最佳会所设计

The Nomination for Best Design Award of Club

Yuan Lu Club

设计者 | Designer

成志、胡俊峰
Cheng Zhi, Hu Junfeng

设计说明 | Design Illustration

搭建一个"南方的院子",让人们更好地去享受自然,是对于这幢建筑和室内设计的理念。最大限度地打开人与自然的关系,让阳光、雨水、落叶、空气、鸟儿……透过院子上方的"开口",没有遮挡地靠近我们。"江南水景"贯穿于建筑始终,庭院布局让人更亲近自然,就地挖掘的淤泥,深加工后完成了建筑所有立面,而温暖和有故事的旧木则是室内的主角。

With the aim to build a "Southern House", letting people hear can have the chance to enjoy the nature, the designer regards this aim as the whole design concept. Make the close relationship between human and nature, including the sunshine, rain, leaves, air, birds... We can feel all from the top of the roof. The rhythm of "Jiangnan waterscape" is full of the entire space, the layout of "courtyard" makes us have chance to contact the nature, and the silt dug with complete processing and used in all facades of the building, the old wood with story is the lead of the interior space.

提名奖
最佳会所设计
The Nomination for Best Design Award of Club

Xuzhou Yunlonghui Clubs
徐州云龙会会所

设计者 | Designer

徐明 Xu Ming

设计说明 | Design Illustration

没有用某个特定的设计理念来框定整个会所的构建。云龙会地处凝聚了千年文化的历史古城，因此将中式的神韵融入了整个空间的设计中。同时，为了符合会所功能的需求，设计导向自然是整个空间餐饮氛围的营造。两者结合，即成了如今所见的，既有中国传统元素，又不那么中式，呈现出现代中式的风貌。

There is no specific design concept for the entire club.The Club is located in a historic geopolitical cohesion with thousand years of culture,so the designer takes Chinese style into the entire space. Meanwhile,in order to meet the functional needs of the club,the main task is to create a perfect dining space.The combination of Chinese traditional elements and modern dining space elements show such a modern Chinese style.

最佳**办公空间**设计奖
BEST DESIGN AWARD
OF OFFICE SPACE

最佳办公空间设计艾特奖 Best Design Idea–Tops Award of Office Space
屈慧颖、冉旭

最佳办公空间设计提名奖 The Nomination for Best Design Award of Office Space
Ippolito Fleitz Group GmbH / 福州名宿装饰设计工程有限公司 / 开山设计顾问有限公司 / 北京艾迪尔建筑装饰工程有限公司

艾特奖

最佳办公空间设计
Best Design Idea-Tops Award of Office Space

ANNND Office
ANNND 共和办公室

设计者 | Designer

Qu Huiying, Ran Xu
屈慧颖、冉旭

设计说明 | Design Illustration

本设计让VI与空间进行了无缝连接，无论接触到抽象的平面还是具象的空间，都穿梭于一个整体而富有感染力的环境当中，在不同的维度、不同的位置都听到属于共和的声音。小空间的设计，赋予空间不同的功能是关键，对空间的利用和再利用成为可能，酒吧、会议、培训和休闲空间的整合是设计的亮点。在设计选材上力求质朴、经济和环保，乳胶漆、水泥、玻璃大面积的使用，用平凡的材料创造出不平庸的效果。

The most efficient convey of this design is the seamless connection between VI and space.The designer wants to give the audience a whole and infectious environment both in an abstract plane and representational space,hearing the sound of the republic in different dimensions and different position. This little space design gives one space different function is the key point.It is possible for the space utilization and reuse;the bars,conference,training space and the integration of recreational space design are bright spots.Spending less,economic and environmental protection were considered in design material selection,emulsion paint,cement,glass used in large area,we hope that we use ordinary materials to create the extraordinary effect.

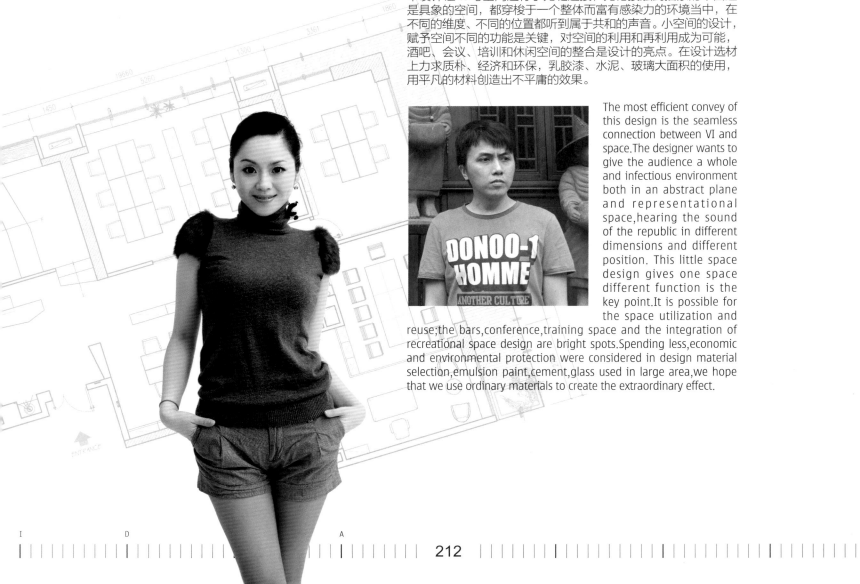

这是一个多功能的办公空间，它的装饰非常朴素：由蓝、灰、黑、白四色，玻璃，树枝，着蜡且抛光的混凝土等材料构造而成。但是，在某些时刻它可以让我们产生某种奇异感。

——Jean-Paul CASSULO 让·保罗·卡苏洛
（法国国家建筑师行会"普罗旺斯-阿尔卑斯-蓝色海岸"行政大区主席）

提名奖

The Nomination for Best Design Award of Office Space

最佳办公空间设计

Drees & Sommer Stuttgart

设计者 | Designer

Ippolito Fleitz Group GmbH

设计说明 | Design Illustration

Drees & Sommer 股份公司新总部办公室彻底完成了公司内部办公思想的转轨，打破僵化的工作岗位，注入"无疆界办公室"的概念，满足了现代化办公方式的要求。灵活的工作和出勤时间，团队规模随时变化，都成为工作中的可能。全新的办公领域由流畅的空间组成，自由设置的半封闭"岛屿"放置了半高的功能性办公家具，将空间划分成区域 。一个视野开敞的办公空间令人感觉与整体空间保持一致。

The redesign of the Drees & Sommer AG headquarters implements the company's shift toward a new working philosophy.The move away from fixed workstations to a "non-territorial office" fulfils the demands of today's working pattern,in which flexible working hours and in-house attendance as well as fluctuating team sizes have become a matter of course.The new office district features flowing spaces zoned by freely positioned island retreats and waist-height functional elements.The offices,separate yet wholly transparent,are perceived as an integral part of the overall space.

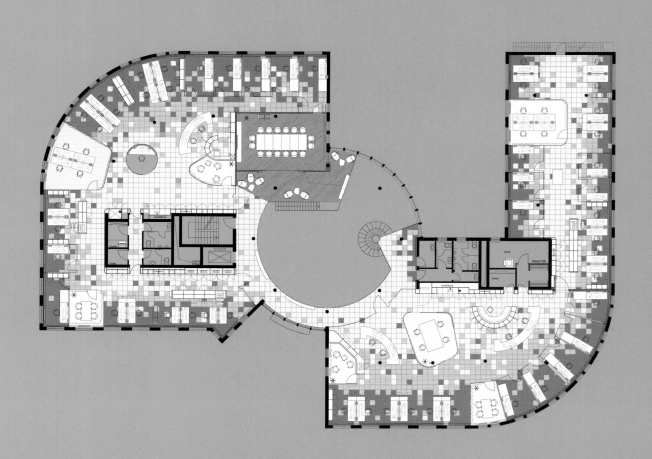

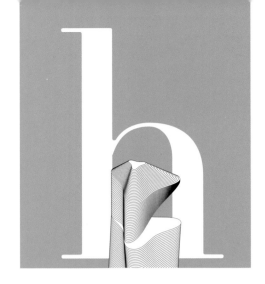

提名奖

最佳办公空间设计
The Nomination for Best Design Award of Office Space

Global Architecture
环宇建筑

设计者 | Designer

Fuzhou Mingsu Decoration Design Engineering Co.,Ltd
福州名宿装饰设计工程有限公司

设计说明 | Design Illustration

本案设在20世纪中期的建筑空间中，我们在保持原建筑工业美感的同时使用天然材质赋予它新的生命。大面积的栽培草平衡了空气中的湿度，天然木纹带来了质朴感，休闲活动的引入让工作氛围更为活泼。这种简洁、环保、轻松的环境给工作带来的是积极有效的动力。

This design case is set in the architectural space in the middle of the last century.While maintaining the industrial beauty of the original building,we use natural materials to give new life to the building. Large areas of cultivated grass balance the humidity and temperature of the air;the natural wood grain brings a rustic sense.Moreover,the introduction of leisure activities makes the working atmosphere more lively.This simple,environmental and relaxed environment provides a proactive and effective power to work.

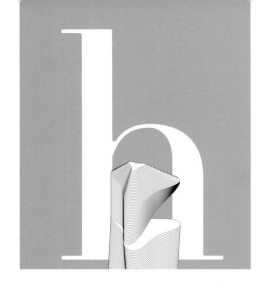

提名奖
The Nomination for Best Design Award of Office Space
最佳办公空间设计

Kaishan Design Consultant Office
开山设计顾问办公室

设计者 | Designer

Kaishan Design Consultant Co.,Ltd
开山设计顾问有限公司

设计说明 | Design Illustration

本案保留了原有老厂房粗墙和老物件的同时加入两个新建筑盒子，盒子的开口能很好地把室外的光线引入工作区域，同时亦对应内部通道。木、石、铁、泥沙，高、低、宽、窄、长、短等设计元素的拆字再组合。新与旧、粗与细在空间中既协调又各具魅力。

This design has kept the brick wall and old articles of the old workshop,meanwhile two new constructing boxes has been added,the opening of these two boxes can guide the outdoor light into the working area well,at the same time,they can correspond with the destructuralization and reconfiguration of design elements of inner passage like wood,stone, iron,silt,height,low,width,length,short,etc.New and old,rough and fine these elements can not only coordinate with each other perfectly,but also has each one's charming.

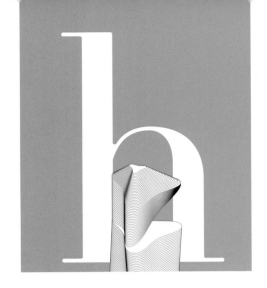

提名奖
最佳办公空间设计
The Nomination for Best Design Award of Office Space

Kaixinmahua Headquarter Office
开心麻花办公总部

设计者 | Designer

Beijing Aidier Building Decoration Engineering Co.,Ltd
北京艾迪尔建筑装饰工程有限公司

设计说明 | Design Illustration

整个设计过程中包含了对老建筑历史的尊重，对新旧建筑协调性的考量以及对主题性室内空间的探索。位于两栋旧厂房中间的新建筑，采用和厂房相同的尖顶造型和一致高度。通过钢锈板材质的带状造型将两栋旧建筑连接起来，优雅对称而个性鲜明。

南侧厂房的办公空间内，延续钢木组合桁架的形式和色彩作为设计主旋律搭建了二层空间，穿过斑驳红砖墙后的楼梯可上至二层。屋顶新开放的天窗将阳光引入室内，阳光下入口前厅处种植了一颗巨大榕树，树荫下设置了三处半开放洽谈空间，由经过特殊处理的纸板材料构筑而成的半圆屏风可自由滑动围合，保证了一定的私密性。

The whole design process shows great respect to the old architectural history,and the consideration of the coordination of new and old buildings with the exploration of thematic polysemy interior space.The new building located at the two old factory building,use the same model and same height of the workshop, connecting two old building through with steel rust planks banded modelling,elegant symmetry and distinctive personality.

South space of building office,continuation of steel composite truss form and color as a design theme of the second floor space building,go through the mottled red brick wall after the stairs can upstair to the second floor.Newly opened roof skylight introduces indoor sunshine,the sunshine lobby entrance area planted a giant banyan,there are three and a half open discussion space under the tree,the cardboard material made special processing into a semicircle screen freely sliding surround close,ensure the privacy.

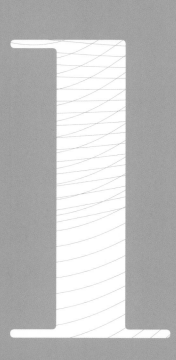

最佳**样板房**设计奖
BEST DESIGN AWARD
OF SHOW FLAT

最佳样板房设计艾特奖 Best Design Idea–Tops Award of Show Flat
冯雷

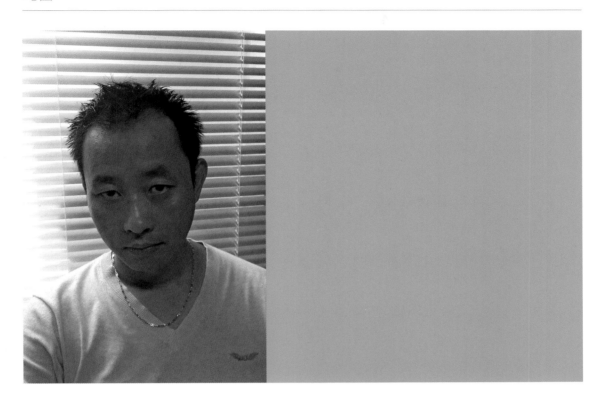

最佳样板房设计提名奖 The Nomination for Best Design Award of Show Flat
王芝密 / 广州市柏舍装饰设计有限公司 / 深圳市盘石室内设计有限公司、吴文粒设计师事务所 / 贾红峰

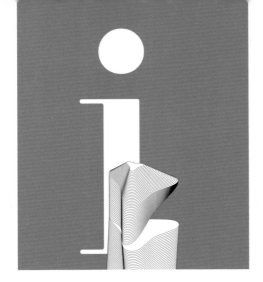

艾特奖
最佳样板房设计
Best Design Idea-Tops Award of Show Flat

Guangyuan Tianyue Mansion

广元天悦府

设计者 | Designer

冯雷 Feng Lei

设计说明 | Design Illustration

在本案设计思路扬长避短的主导思想下，重室内、弱室外。室内分区以结构柱为分割核心，采用对称化、对应性的处理方式，划分空间主次功能，并以空间使用与功能的主次流线，结合建筑采光依次设置空间使用功能。同时通过空间块面凹凸造型，利用材质本身色泽、质感属性，结合平面功能区划，以转折、围合的手法区分功能。

In this case, the designer pays much attention to the interior space but the exterior space, and takes the structural column as the interior partitions, makes use of the symmetry of the correspondence approach, dividing the space of primary and secondary functions, distinguishes the use of space and function of primary and secondary flow lines, combines with the use of architectural lighting in order to set the function space. Meanwhile the designers make the concavo convex shape through block space, and the use of the material itself, color, texture attributes combine functional areas zoned plane to turn, enclosed way to distinguish features.

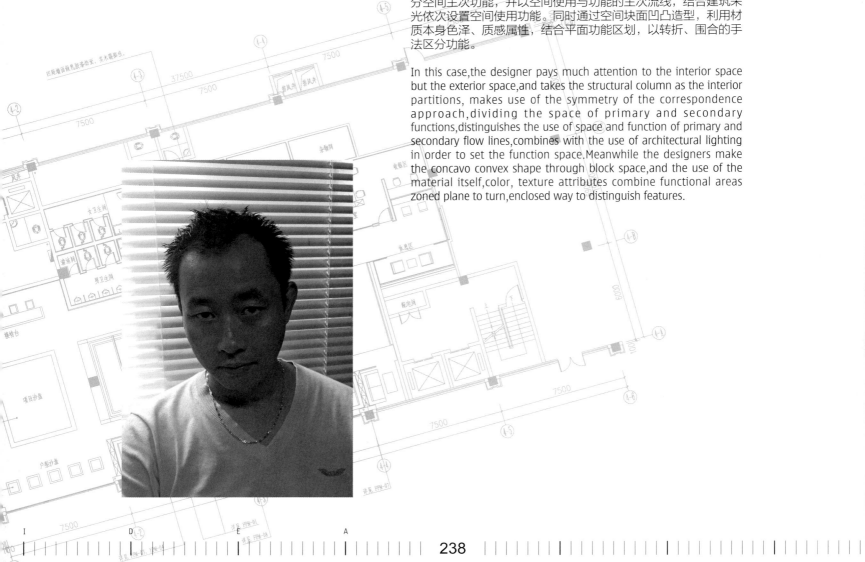

Functional elegance, balanced selection of materials and lighting. 优雅的功能设计,独特的设计构思,项目设计实现了材料与光线的平衡。

——Ilhan Zeybekoglu 依鲁翰·泽碧克格鲁
(美国 ZNA 泽碧克建筑设计事务所创始人、美国哈佛大学建筑学院教授)

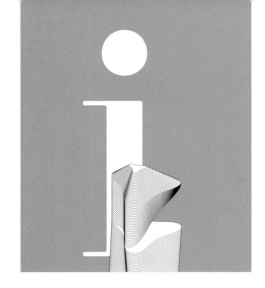

提名奖

最佳样板房设计
The Nomination for Best Design Award of Show Flat

Xiamen Yongjing Bay Show Flat
厦门融景湾样板房

设计者 | Designer

王芝密 Wang Zhimi

设计说明 | Design Illustration

极简、内敛的黑与白，加上灰色、暮色，低调的中性色带出独一无二的空间品位。东方的空间观念和禅学隐而不露地融入整个设计中，让现代文化之家展现致远的东方表情，有着书香门第的文化语境。

The simple Chinese style show flat design—fashion,culture and taste,the minimalist,restrained black and white color, with gray,wood color,low-key space make a unique ribbon neutral taste.Oriental Zen concept of space merges with concealing Zen spirit overall design,so that the modern culture house shows a quiet oriental atmosphere,with scholarly cultural context.

提名奖

最佳样板房设计
The Nomination for Best Design Award of Show Flat

Dongguan Dragon Spring Yuan Show Flat
东莞龙泉豪苑样板房

设计者 | Designer

Guangzhou Percept Space Interior Design Co., Ltd.
广州市柏舍装饰设计有限公司

设计说明 | Design Illustration

本项目空间布局与设施都与酒店相仿，配合各种造型的水晶灯饰，营造一个现代华贵的居住空间。客厅天花吊顶水晶灯，由干邑色的水晶珠串联，构成三个指环相互交错，整体用玫瑰金镜钢包边，工艺考究。卧房气氛以紫色调为主，通过屏风夹丝玻璃连接书房，手绘墙纸搭配紫色的沙发、窗帘，手织的植物图案地毯，大气沉稳。玫瑰金镜钢的多处装饰倾吐芳华，花鸟手绘墙纸精致而充满味道，点缀其间的几个立体感抱枕，从细节感受整个空间温暖华丽的神韵。

The layout and facilities of this project is similar with hotel,with a variety of modeling crystal lighting, creates a modern and luxurious living space.The ceiling crystal lamp of living room,crystal beads cognac colored by a series of three rings intertwined constitute the whole mirror steel with rose gold edging,sophisticated technology.Bedroom with purple tones dominate the atmosphere,using the wire glass screen connects the study room,hand-painted wallpaper with purple couch curtains,hand-woven carpet pattern plants to create a calm atmosphere.Various decorative steel rose gold mirrors,delicate and full of birds and flowers hand-painted wallpaper taste,a few three-dimensional embellishment pillows make the entire space feel warm and gorgeous charm from the details.

提名奖
最佳样板房设计
The Nomination for Best Design Award of Show Flat

Hubei Yichang Hengxin Central Park 9A Show Flat
湖北宜昌恒信中央公园 9A 样板房

设计者 | Designer

Shenzhen Huge Rock Interior Design Co., Ltd
Wu Wenli Design Firm
深圳市盘石室内设计有限公司
吴文粒设计师事务所

设计说明 | Design Illustration

蓝色的纯净总是让我们想起广阔的天空和静谧的海洋，它不仅带给人们轻盈的步调，且能令人感觉到丝丝惬意。利用局部的蓝色软装配饰作为点缀，采用暖色调中和整体色感，打造清新宜人的居室氛围。条纹、花案、描边，这些不同图案的床品和靠包混搭在一起，为空间增加深度和冲击力，在营造出时尚奢华感的同时不失清新盎然。

Blue purity always reminds us of the vast sky and the silence of the sea,which not only lets us have light pace,but provides a comfortable space.The designer uses blue to decorate and to be the embellishment,and use the warm colors to integrate the overall sense of color,to create fresh and pleasant room atmosphere.Stripes,floral and stroke,all these different patterned beddings and back cushions mix together for space to add depth and impact,creating a sense of stylish luxury and making a fresh and full of life.

提名奖

最佳样板房设计
The Nomination for Best Design Award of Show Flat

The Fifth Garden
第五园

设计者 | Designer

贾红峰 Jia Hongfeng

设计说明 | Design Illustration

以一栋老房子为基础来改造，采用东方与西方有机结合的手法，定位于高端人群的文化品鉴追求。在环境风格上以现代中式语言的建筑环境和古典中式建筑结合古典西式家具风格，凸显出中西文化的意味传达。空间布局力求四平八稳对称式布置，让您坐享传统乐趣，品味西式意韵。

The Fifth Garden is a renovation project,combines the elements of east and west,services for the high-end group, who pursuit for the culture tasting.The style of modern Chinese language environment and classical Chinese architecture combines classical Western style furniture,to highlight the soul of Chinese and Western cultures.Spatial layout strives cautiously symmetrical,allowing you to reap the traditional fun, taste Western rhyme.

最佳**别墅豪宅**设计奖
BEST DESIGN AWARD
OF VILLA

最佳别墅豪宅设计艾特奖 Best Design Idea-Tops Award of Villa
Bruno Erpicum

最佳别墅豪宅设计提名奖 The Nomination for Best Design Award of Villa

Whitebox Architects,Panagiotis kokkalidis,Aliki TrianTafitlidoy,Anna drella / Rolf Ockert / 深圳市矩阵室内装饰设计有限公司 / 上海亚邑室内设计有限公司

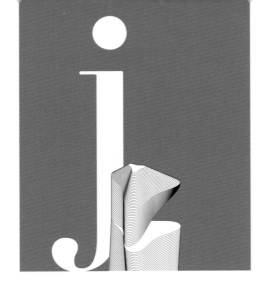

艾特奖

Best Design Idea-Tops Award of Villa
最佳别墅豪宅设计

AIBS

设计者 | Designer

Bruno Erpicum

设计说明 | Design Illustration

恰如一条尽头被封闭的小路或者小径，地面被一个楼梯打开，穿越楼梯到桥上可以欣赏这里所有的美丽景观。生活区被一个大的窗户包围，这面窗户同时可以防风。沿着封闭的庭院区域也有一扇大窗户。悬崖顶部有一棵橄榄树提供给庭院第二层保护。视线较远的地方，游泳池被周围的自然环境所包围，位于超越露台建筑的侧面，很多墙壁和柱子都被精心建于混凝土表面之上，并支持地面的空间结构，其中包含卧室。位于159米的高空，没有任何建筑物的功能构成了对大自然的威胁。蓝天下的建筑显得平静和安详，而在暴风雨天气它又显现出了引人注目的压抑气氛。

Just like a path or road which comes to a dead end,the land becomes rippled before turning into a staircase which leads you down to the lower bridge from where you can appreciate the landscape in all its beauty.The living areas are enclosed by a single large window frame. The windows also provide protection against the winds.There are also large windows along the patio which is in an enclosed area. The cliff which has an olive tree on top provides a second wall for the patio.Away from view,the swimming pool lies to the side of the building beyond the terrace,surrounded by the natural environment. A number of walls and pillars have been painstakingly erected on the concrete surface and support the floor above which contains the bedrooms.Located at 159 metres altitude,none of the building's features constitute a threat to nature.Under blue skies the building appears calm and serene whilst in stormy weather it has a striking and tormented air about it.

项目胜出的优势在于充分利用地形的所处环境,将建筑的内外空间融为一体。室内的观景视线极佳,采光与照明的设计恰到好处。材料与色彩的运用与建筑风格相得益彰,陈设物的选择体现了少而精的原则,营造出高雅的环境氛围。

——郑曙旸

(国务院学位委员会设计学学科评议组组长、清华大学美术学院教授、博士生导师)

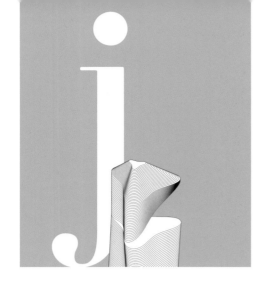

提名奖
The Nomination for Best Design Award of Villa
最佳别墅豪宅设计

Stone House in Greece

设计者 | Designer

Whitebox Architects,
Panagiotis kokkalidis,
Aliki TrianTafitlidoy,
Anna drella

设计说明 | Design Illustration

这是一个四口之家的住所，家里有一对夫妻及两个小孩，设计时需要考虑留一间客房，并匹配有自主独立的卫浴。其基本要求是：四间卧室均可观海景，出于工作的考虑一楼需要安排工作区域，但大多是为了母亲而设计，以便于她一边工作一边监督玩耍的孩子们。另一个要求是房屋设计尽量环保节能，并且可以全年享受部分室外空间，包括餐饮、游泳、娱乐。

The concept was the creation of a residence for a family of four—the parents with two children—and the possibility of having a guest room with relative autonomy—separate bathroom.The basic demands were:the view of the sea from all four bedrooms,an office space on the ground floor for the professional needs of the couple but mostly of the mother who wanted to work and supervise the ground floor where the children would play.Another request for the design was the economy in energy consumption of the house and the possibility of enjoying the outdoor spaces throughout the year,for dining,swimming,games.

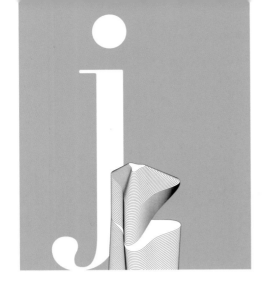

提名奖
最佳别墅豪宅设计
The Nomination for Best Design Award of Villa

Bronte House

设计者 | Designer

Rolf Ockert

设计说明 | Design Illustration

客户希望我们可以创造出他们梦想中的家，就如同倚靠太平洋海岸的一个乐园。在这里可以让人感受到仿佛每天都在度假。这是一个新颖独特的构思，然而房间的占地面积并不大，并且周围的居住环境比较拥挤。

The client approached us to create house of their dreams on a site perched high over the Pacific Ocean,a home that was to make them feel like being on holiday every day.While the view was fantastic, the site was very small and suffocated by overbearing neighbouring dwellings.

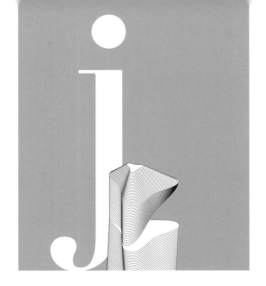

提名奖
最佳别墅豪宅设计
The Nomination for Best Design Award of Villa

重庆万科城别墅

设计者 | Designer

Shenzhen Matrix Interior Decoration Design Co., Ltd
深圳市矩阵室内装饰设计有限公司

设计说明 | Design Illustration

大面积深色木衬底没有过多的精雕细琢，更多的都是通过两到三种或光滑或粗犷相对比较强烈的材料映衬，将传统意境和现代风格融合运用，用现代设计手法将东方的语汇去繁从简地表达出来。没有多余的造型雕花、纯粹的深色水曲柳饰面、浅色麻面墙布的搭配既传统又流行，而且为空间营造出了充满魅力的对称感，使整个空间更具立体感，在美观之余，更增韵味，彰显东方的气质。

Large area of dark wooden substrate is left blank, instead of exquisite carving, most of the area is filled by soft or rough materials with sharp contrast of light and shade. We balanced the traditional artistic conception with modern style, expressing oriental conception through modern design method. With no needless carvings, the match of Chinese White Ash screen with Dark Granite is traditional but popular, which creates a glamorous spacial symmetry, showing a more stereoscopical space not only beautiful but rather charming with the unique oriental classic elegance.

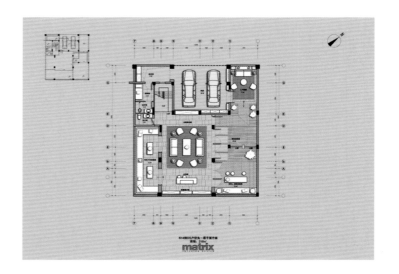

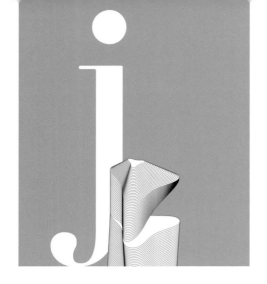

提名奖
最佳别墅豪宅设计
The Nomination for Best Design Award of Villa

Shanghai Pujiang OCT 122-10Villa

浦江华侨城 122-10 地块别墅

设计者 | Designer

Shanghai Yayi Interior Design Co.,Ltd

上海亚邑室内设计有限公司

设计说明 | Design Illustration

本案力图展现一派奢华禅意的风格。当光和影在空间交错更迭，空间随着视线的角度变化，变得宁静而自然。日光毫无遮掩地穿过窗框和玻璃照了进来，华丽感从客厅的石墙上穿越，停在客厅金属与粗犷石材相互融合的装饰墙上，这就是开阔的厅堂所带来的视觉盛宴。

This case is a rich elegant breath of independent villas,showing a posh zen style.Temperament is convey through the eyes of heart throb.When light and shade staggered change in space,with the line of sight Angle change,it becomes quiet and natural,sunshine come in through the window frames and glass,luxuriant sense through from the wall of the sitting room,stop the adornment of the metal in the sitting room accommodation with rugged stone wall,this is tall sharing,open hall brings the visual feast.Nature's charm,refine out with contemporary gimmick.

最佳陈设艺术设计艾特奖 Best Design Idea–Tops Award of Art Display
LIKEarchitects

最佳陈设艺术设计提名奖 The Nomination for Best Design Award of Art Display
天坊室内计划有限公司 / 葛亚曦 / 深圳市天琢轩设计有限公司 / 维斯林室内建筑设计有限公司

艾特奖 Best Design Idea-Tops
最佳陈设艺术设计 Award of Art Display

X – HIBITION

设计者 | Designer

LIKEarchitects

设计说明 | Design Illustration

为一个250平米的空间进行空间设计，这个空间将承担两个不同的大型活动。其一，20个技术初创企业的展览；其二，100人左右的会议。因此，要求设计区域高度灵活，可以在短短几个小时内完成空间结构功能的转化。

The challenge was to design space partitions for a 250 sqm room that would host two different programs on consecutive days:on the first,an exhibition on twenty technological start-ups;on the second,a conference for 100 people.Thus,the proposed intervention should be highly flexible and able to transform the spatial qualities of the room in just a few hours.

在室内设计中陈设艺术必须是以画龙点睛的作用呈现,并成为空间的主体控制物,才具备空间艺术氛围营造的效果。该项目正是达到了这样的标准:空间造型的选择精确到位,尺度与比例的推敲恰当适度,色彩与光的环境运用,准确诠释了空间场所的概念主题。

——郑曙旸
(国务院学位委员会设计学学科评议组组长、清华大学美术学院教授、博士生导师)

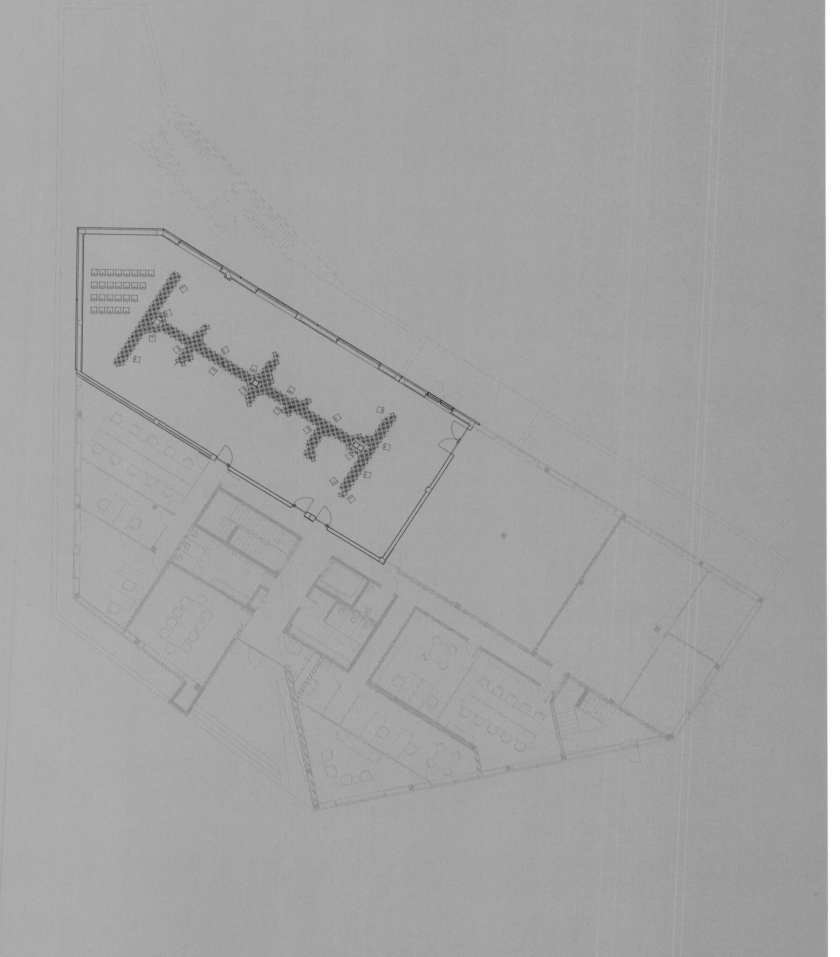

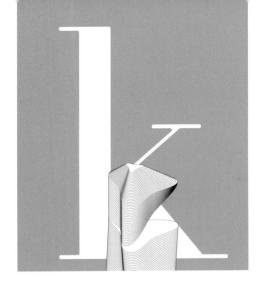

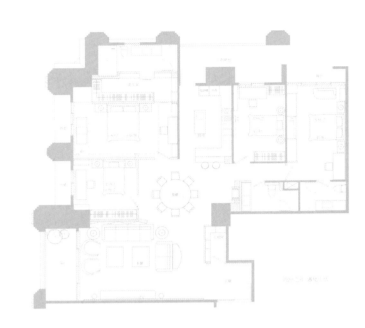

提名奖

The Nomination for Best Design
Award of Art Display
最佳陈设艺术设计

Live a Life Outside the Design
设计之外 遇见生活

设计者 | Designer

Tian Fang Interior Planning Co.,Ltd
天坊室内计划有限公司

设计说明 | Design Illustration

流畅协调的格局实现了一种可视化的干净，这干净饱满而丰富，本身即是一种独特的设计，它让背景不仅仅是背景，成功地烘托出了家具和室内装饰品的质感，同时也饱含了对历史以及文化的尊重与追求。

Smooth coordination pattern implements a visualization of clean,this is a kind of unique design itself.It is not only a background,oil out the simple sense of furniture and interior decoration successfully,but also show respect and pursuit to the history and culture.

提名奖

最佳陈设艺术设计

The Nomination for Best Design Award of Art Display

Shanghai Greentown Yulan Garden
上海绿城玉兰花园

设计者 | Designer

葛亚曦 Ge Yaxi

设计说明 | Design Illustration

在现代简约、精致严谨的硬装基础上，LSDCASA 以新禅意的手法将东方的"神"与西方的"形"碰撞产生对话，具象地展现了更具维度和深度的空间美学，为品味艺术、喜爱收藏的主人量身打造符合生活与收藏习惯、散发东方智慧的新高度住宅。

The 257.5 ㎡ prototype room a total of four layers,a layer, layer,respectively,the basement and underground mezzanine. In contemporary and contracted,refined based on rigorous hard outfit,LSDCASA like new zen in the "god" in the east and the west "shape" collision dialogue,representational showed more dimensions and depth of space aesthetics,for the taste of art,collectors master tailor conforms to life and habits,reveal a new height of the Oriental wisdom.

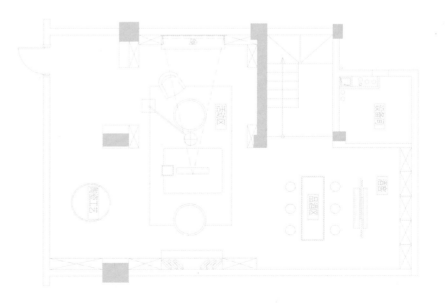

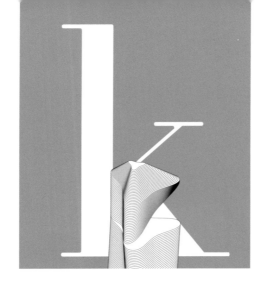

提名奖
最佳陈设艺术设计
The Nomination for Best Design Award of Art Display

Mei Houssi
美豪斯

设计者 | Designer

Shenzhen Tian Zhu Xuan Design Co.,Ltd
深圳市天琢轩设计有限公司

设计说明 | Design Illustration

本案所处的空间公共走道狭窄，为使其商业价值最大化，引用"退一步海阔天空"的思维方式，而做出"以退为进"的设计理念，使其空间退后一大步，并加一面巨大的造型镜子，使空间变得无限开阔深远，使公共空间和本案空间相融合。

This case is a space of narrow public walkway, to maximize its business value, refer to "step back to get higher" way of thinking, take the design concept, make the space back a step, and the modelling of a giant mirror, make the space becomes infinite far-reaching, let the case integrate to the open public space.

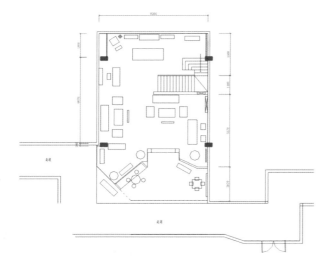

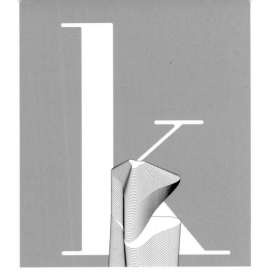

提名奖
最佳陈设艺术设计

The Nomination for Best Design Award of Art Display

The Joyful Tree House

设计者 | Designer

维斯林室内建筑设计有限公司
Wei Si Lin Interior Architectural Design Co., Ltd

设计说明 | Design Illustration

半山一号的精髓就是提升了室内设计的质量门槛，流露并散发出标新立异的意义，启发于树木的灵感，设计图形丰富地融入了空间，给住户们带来自然平和之感。

The essence of mid-levels NO.1 is improvement in the quality of threshold which exist in the interior design industry, reveal and sends out a different meaning. Inspired by trees of inspiration, rich patterns into the space, brings residents natural and peaceful feeling.

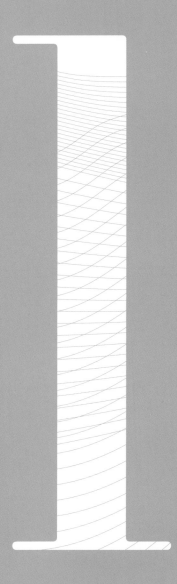

1

最佳**公寓**设计奖
**BEST DESIGN AWARD
OF APARTMENT**

最佳公寓设计艾特奖 Best Design Idea—Tops Award of Apartment
Susanna Cots

最佳公寓设计提名奖 The Nomination for Best Design Award of Apartment
黄仁辉 / 国广一叶装饰机构 / 隐巷设计顾问有限公司 / 杰玛室内设计有限公司

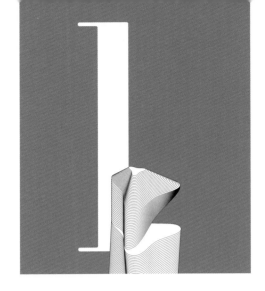

艾特奖

Best Design Idea-Tops
Award of Apartment
最佳公寓设计

Transversal expression

设计者 | Designer

Susanna Cots

设计说明 | Design Illustration

该项目用木质传递设计师的设计理念，设计师将这一理念淋漓尽致地体现在了这座巴塞罗那两位读书爱好者的家中。整个项目中，在同一层楼的整个空间大动脉是木漆板，跨越到客房中，并营造出一种温暖的保护氛围，木板材质在空间中垂直和水平地穿插创造出一种情感和温暖的通道。

A skin made of wood that transversally crosses the housing is Susanna Cots, "leitmotiv" to develop this interior design project,the home in Barcelona of two book lovers. On the same floor,the main artery that unites the whole space is a wooden lacquered sheet that crosses the rooms and wraps them up like a warm and protective skin.Wood crosses the apartment vertically and horizontally creating an emotional and warm passageway.

A simple and elegant creation of common living spaces, and intelligent use of materials. 设计师打造了一个简单优雅的普通生活空间，对材料的选择也非常睿智。

——Ilhan Zeybekoglu 依鲁翰·泽碧克格鲁
（美国 ZNA 泽碧克建筑设计事务所创始人、美国哈佛大学建筑学院教授）

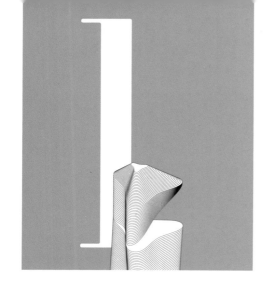

提名奖
The Nomination for Best Design Award of Apartment
最佳公寓设计

Seeing Beautiful Horizion

看见·美丽的平行线

设计者 | Designer

黄仁辉 Huang Renhui

设计说明 | Design Illustration

"捷运线"孕育一条城市规律生活的"平行线",本案借由"看见·美丽的并行线"来隐喻一种享受城市生活与空间美学的连结,将室内空间的线条抽象"平行"于城市的"捷运线",并运用连续的材质构成空间力量,借由材料分割转换成空间的线条韵律,将情感和记忆的并行线收藏在丰富的立面表情内。

"Express line" develops a rule of life for the city's "parallel lines";The case—Seeing Beautiful Horizon shows the links of enjoying the city life and space aesthetics in a metaphorical way,taking the abstract lines of the interior space parallels the urban express line and using consecutive material to make the space power,transforming the separated material into lines rhythm.The emotions and memories were collected in the rich facade faces.

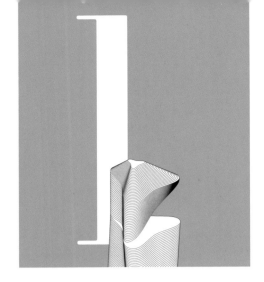

提名奖
The Nomination for Best Design Award of Apartment
最佳公寓设计

The Book Rhyme
书韵

设计者 | Designer

Guo Guang Yi Ye Decoration Group
国广一叶装饰机构

设计说明 | Design Illustration

本案以新东方的表现形式营造书香古朴的雅士情调，同时注重审美与实用功能恰到好处的结合，不失时尚简约的现代生活意趣。传统与现代，古典与时尚，艺术和日常，内心与外物，在这个空间里再一次寻到了相互杂糅、融合、交相辉映的可能。

This project,with the neo-orientalism manifestation,creates a literary ancient poets atmosphere,focusing on aesthetic and functional proper combination,showing chic,modern and charm life.Traditional and contemporary,classic and fashion,the sense of art and daily life,heart and matter finds the possibility of integration and amalgamation in this space.

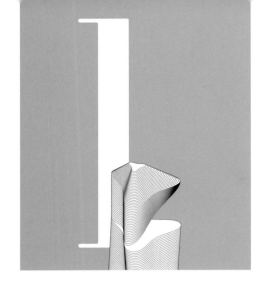

提名奖
最佳公寓设计
The Nomination for Best Design Award of Apartment

National Art Galaxy of Guo
国家美术馆郭邸设计案

设计者 | Designer

XYI-DESIGN CO.,LTD
隐巷设计顾问有限公司

设计说明 | Design Illustration

"光线、动线与质感"为本宅的设计理念。天花板的设计主要是保持空间高度，并以全局照明的概念处理。入口玄关与餐厅的天花板是空间的最低点，玄关地面为倒角实木地板，强化脚踩的感受。进入客厅之后空间感因天花与木格栅产生变化，而地面也转换成大理石，这点是考虑人从外面嘈杂的环境回到家中，第一个感受应该是平静、稳定感，随着不同空间与光影变换，让心理慢慢放松，天花上的银河状光纤则是体现业主的浪漫个性。

The designer takes "light, design lines and texture" as the design concept of the space, the ceiling is designed primarily to keep space height, to create a concept of global illumination, and from the space, you can see the ceiling of entrance hallway and restaurant is the lowest point in the space; the ground of the hallway is chamfer solid wood flooring, which strengthens the feelings of touching; when you step into the living room, the sense of space is changed by the ceiling and wooden grille, and the ground is also converted into marble; the reason of such a design, the designer is considering that the owner come from a noisy environment, the calm and stable sense of space is needed; the light of this space with different transformations, making psychological slowly relax, and the owners' romantic personality is reflected by the galaxy shaped fiber of the ceiling.

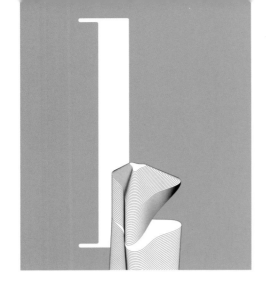

提名奖
The Nomination for Best Design Award of Apartment
最佳公寓设计

Relax Meditation
松·定境

设计者 | Designer

Chongqing Jiema Decoration &Design Co., Ltd
杰玛室内设计有限公司

设计说明 | Design Illustration

在一入门的玄关，我们将屋主长辈赠与的佛首收藏摆设于此，利用颜色和材质搭配，呈现出沉静的意象，也作为一个转换心境的空间。房屋本身的天花板低，加上大柱子十字梁经过，因此设计师运用造型修饰的手法，在天花板做折板和斜面设计，搭配光源增加视觉层次，产生高低的错落感，来减少视觉上的压迫。并利用梁柱的特性，在空间中形成隐形隔间，区分出了客厅、开放式书房、餐厅与厨房。

The designer puts the Buddha figure, gave by the eldership for the homeowner in the entrance hallway,and shows a quiet imagery and a mind change space by matching color and texture. Because of the low ceiling, large pillars and rood beam, the designer uses model modified approach to do flap and bevel design in the ceiling;increasing the visual level with light to develop a scattered sense of high and low and reducing the oppression of the visual.The special characteristic of beams forms invisible compartments in space,which distinguishes the living room,an open study,dining room and kitchen.

最佳**商业空间**设计奖
BEST DESIGN AWARD
OF COMMERCIAL SPACE

最佳商业空间设计艾特奖 Best Design Idea–Tops Award of Commercial Space
张海龙

最佳商业空间设计提名奖 The Nomination for Best Design Award of Commercial Space

厦门一亩梁田装饰设计工程有限公司 / 广州市五加二装饰设计有限公司 / 上海明合文吉建筑设计有限公司 / 广州三木鱼建筑设计咨询有限公司

艾特奖

最佳商业空间设计
Best Design Idea-Tops
Award of Commercial Space

Yuzhouyunding International
禹洲云顶国际

设计者 | Designer

张海龙 Zhang Hailong

设计说明 | Design Illustration

也许有时候，往往简单的手法却能表达一些复杂的思想，卸下枷锁回到设计师的初衷。本案无论是错落有序的木隔断还是洁白粗糙的倾斜墙都是棱角分明的直线条，加之沉着冷静的色彩搭配，表达出设计师成熟大胆的设计思路。然而洁白和亮黄的灯光，无疑是给整个空间带来别样生机。

Maybe sometimes,simple way can express complex ideas, break free back to the designer's original intention.The case whether the wooden partition of the staggered or rough inclined wall are angular white straight line,together with the color collocation of the calm,expresses the designer mature bold design ideas.Pure white and bright yellow light,however,is undoubtedly bringing another life to the air.

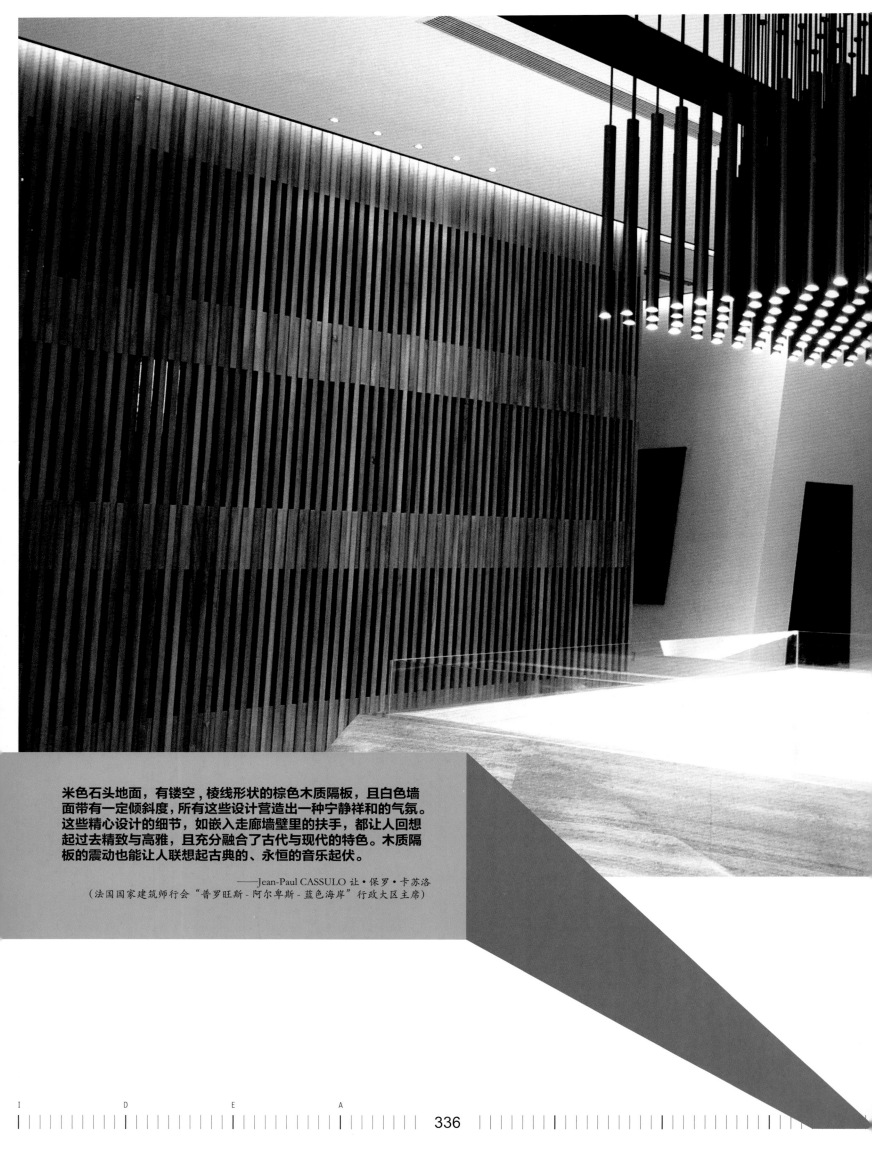

米色石头地面，有镂空，棱线形状的棕色木质隔板，且白色墙面带有一定倾斜度，所有这些设计营造出一种宁静祥和的气氛。这些精心设计的细节，如嵌入走廊墙壁里的扶手，都让人回想起过去精致与高雅，且充分融合了古代与现代的特色。木质隔板的震动也能让人联想起古典的、永恒的音乐起伏。

——Jean-Paul CASSULO 让·保罗·卡苏洛
（法国国家建筑师行会"普罗旺斯-阿尔卑斯-蓝色海岸"行政大区主席）

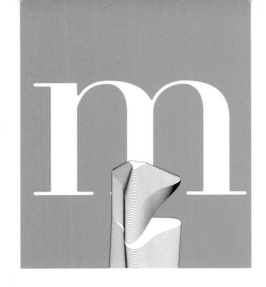

提名奖

最佳商业空间设计

The Nomination for Best Design Award of Commercial Space

Nest Salon
巢沙龙

设计者 | Designer

Yi Mu Liang Tian Decoration Design Engineering Co.,Ltd
厦门一亩梁田装饰设计工程有限公司

设计说明 | Design Illustration

本案地处于厦门繁华路段，纷繁的都市早已掩盖了人们内心那份简单纯真的渴望。设计师摒弃了浮华的表皮，炫目的造型，力求还原一个单纯的空间。通过巧妙运用平板灰水泥瓦，灰色瓷砖，为顾客带来一个轻松简单纯粹的灰色调空间体验。怀旧年代的壁灯，算盘装置拼画以及一些从二手市场淘来的旧家具、石条，为空间营造了富有文化底蕴的气息，提升了整个沙龙的档次和整体气质，展现了单纯的美。

The case located in downtown section of Xiamen,the city already covers the heart of the pure and simple desire.Designers reject ostentatious epidermal,dazzling form, and try to restore a simple space.Through the clever use of flat gray cement tile,grey tiles,to bring customers a simple pure gray space experience.Nostalgia in the wall, the abacus device together and some from the secondary market to get the old furniture,stone,as the space to create a rich cultural heritage.To enhance the entire salon grade and the overall temperament,show the simple beauty.

提名奖

最佳商业空间设计

The Nomination for Best Design Award of Commercial Space

Guangzhou Textile Exhibition Shop
广州纺织博览中心展览空间

设计者 | Designer

Guangzhou Five Plus Two Decoration Design Co.,Ltd
广州市五加二装饰设计有限公司

设计说明 | Design Illustration

天花的变化突出了不同质感的材料，进行了有序组合，给人带来完全不同的视觉感受，整个空间约有6米高。针对不同的布料、款式，使用较高功率的LED灯，从而产生清晰的色彩，很好地强调了展示物品的立体质量，创造了引人注目的商店场景。一个个装置感很强的展架，令产品可以随意组合。布匹展示板的设计令布匹悬挂其中，又为空间增添了些许跳跃的气氛。

Smallpox changes highlighted the different texture of materials between orderly combination,bring people completely different visual perception.Whole space is about 6 metres high,according to different fabrics,style to use higher power LED lights,to produce clear color,emphasized the well stereoscopic quality of display items,create eye-catching store scenario.Each device is exhibited strong frame,make the product can be arbitrarily combination.Cloth display board design makes the cloth hanging,and added jumping atmosphere for the space.

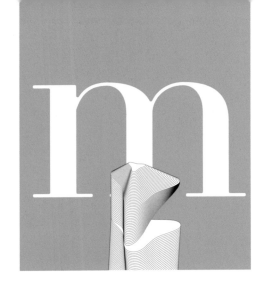

提名奖
最佳商业空间设计
The Nomination for Best Design Award of Commercial Space

Shanghai Bund Hong Kong Duddell Street Flagship Store
上海滩香港都爹利街旗舰店

设计者 | Designer

Shanghai Ming He Wen Ji Architectural Aesign Co.,Ltd
上海明合文吉建筑设计有限公司

设计说明 | Design Illustration

通过设计，改变这种显而易见的东方特色，让人们感受到摩登、优雅的中式设计，让品牌的空间更加摩登，并且使空间更好地衬托产品。在新旗舰店的设计中，设计师通过空间的划分、中式元素的点缀等现代设计手法，带给上海滩更加优雅与现代的气质。通过细节让顾客感受到空间的美妙与品牌的中式内涵，让顾客与空间建立起一种良好的互动。

Design,to change this obvious oriental characteristics,make people feel modern and elegant Chinese style design,make brand space more fashionable,and make a space better foil products. The design of flagship stores,through separated by the space,Chinese style element of the ornament such as modern design methods,to bring more elegance and the temperament of modern Shanghai,through the details to make the customer feel the beauty and the brand connotation of the Chinese style of the space,to make the customer established a good interaction with space.

提名奖

最佳商业空间设计

The Nomination for Best Design Award of Commercial Space

Guangzhou Naturade Slimming & Beauty Showroom
广州奈瑞儿美容品牌形象空间

设计者 | Designer

Guangzhou Mark Fish Architectural Design Consulting Co., Ltd
广州三木鱼建筑设计咨询有限公司

设计说明 | Design Illustration

作为为女子服务的商业空间，设计师主张东方养生、服务女性。通过空间的组织，艺术化地彰显了水之纯朴和东方人文的纯净。选用灰色来作为主要的设计颜色，贯穿整个空间，以东方女性美的造型元素，创作了品牌形象雕塑，与品牌相呼应。灯光的安排，我们强调的是东方的那种宁静。我们努力营造出东方宁静的空间氛围，充分体现一种宁静而简朴的情感。

As a commercial space service for women.Designer advocates eastern healthy style and service women,through the organization of the space,to show the pure water,the purity of the east.Grey is choosen for the main design color,throughout the entire space,with oriental female beautiful modelling element,to create the brand image of the sculpture,which corresponding with the brand.The arrangement of the light, we emphasize that the calm of the east.We strive to build a east peaceful space atmosphere,fully embodies a kind of quiet and simple emotion.

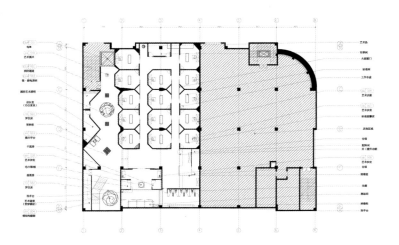

每一个懂艺术的人
都是艺术·家

汇泰龙专注门锁艺术十多年

心灵在忙碌中倦怠，在沉静中裸露出来，
雅舍，一个隐秘优雅的心灵家园，
洗礼着那份脆弱已久的心怀。
在这里，
卸下尘世的喧嚣，
看庭前花开花落，望天空云卷云舒，
那是真诚而惬意的生活。

国际影星 黄圣依
强势代言：

扫描二维码
关注更多精彩

五星级生活典范—汇泰龙

○ 智能门锁产品系统　　○ 五金产品系统　　○ 卫浴产品系统

中式古典系列—雅舍

灵感源于古典建筑中的雕花门窗，精美的雕花表达了古人对于美好居住环境的追求，体现了人与自然和谐共处的境界。

设计灵感源于古典建筑雕花门窗

汇泰龙五金卫浴

全球免费热线：400-880-8333
国际互联网：www.hutlon.com

倫勃朗家居
Rem Brandt Furniture

尊享奢华 品味之选
24K鍍金歐式家具·飾品
24k Gold Plating Furniture And Decoration

钟表系列

饰品系列

灯饰系列

旗舰店： 佛山家居博览城A座一楼1011

商务热线： 86 757-66696822

工厂地址： 佛山市顺德区龙江镇旺岗工业区龙峰大道43号

工厂电话： 86 757-23223083

网站

微信

MAYROSECERAMIC
May Rose
五月玫瑰

- 艾特奖特别贡献奖
- 设计师推荐品牌

五月玫瑰 陶瓷 MAY ROSE

全球顶级设计别墅工程专用陶瓷
THE SPECIAL TILES FOR THE GLOBAL TOP DESIGN VILLA

公众微信二维码

小号微信二维码

玫瑰热线/0757-82562998

地址：佛山市禅城区南庄镇华夏陶瓷博览城陶北路2座7号
www.fsmayrose.com

远鹏·A963

中国 / 深圳 / 福田保税区

国际设计产业园
INTERNATIONAL DESIGN INDUSTRIAL PARK

荟聚高端资源·贯通全产业链·室内设计定制商务平台

招租 **0755-83583007**
热线 13510459250·15814454404

远鹏·A963　地址：中国深圳福田保税区红柳道1号

中国建筑装饰行业百强企业

深圳市建装业集团股份有限公司

深圳市建装业集团本着"团结、求实、进取、卓越"的企业精神，经过十年创业，现已发展成为一家以建筑装饰工程为主，集项目投资、设计研究、体育产业、新材料研发、饰品及家具等多元化经营发展的集团公司。

深圳市建装业集团成立于2001年，是国家建筑装饰双甲级企业，并取得建筑幕墙、建筑智能化、消防、机电设备安装工程设计与施工壹级资质及园林、体育设施、钢结构等专业资质，是中国建筑装饰协会、广东省建筑装饰协会、深圳市装饰行业协会常务理事单位。集团已通过了ISO9001、ISO14001、OHSAS18001国际标准管理体系认证，并连续多年被资信评估机构评为"AAA"级资信企业，且连续多年被工商行政管理部门认定为"守合同、重信用"企业，在业内已树立了良好的企业品牌和形象。

深圳市建装业集团成立以来，一直保持着高速的发展态势，业绩连年来保持成倍数的增长，在全国建筑装饰行业连续多年被评为"百强前20位企业"，同时被评为"明星企业"。此外，集团在装饰工程方面多次获得"鲁班奖"、"全国建筑装饰工程奖"、"各省优质工程奖"、"深圳市金鹏奖"等荣誉，还被评为"广东省企业500强"企业、"亚洲（行业）十大公信力品牌"、"2011-2012年度十大最具影响力设计机构"、"中国建筑装饰行业百强第十七位"、"企业文化建设优秀十佳单位"。近年来，深圳市建装业集团陆续和万科、万达、佳兆业、恒大、保利、富力、雅居乐、碧桂园、珠光、华润、招商、华侨城等一线房地产企业进行了战略合作。这充分表明我集团已全方位迈向国内一流建筑装饰企业的行列，并已成为同行业中的翘楚。

深圳市建装业集团股份有限公司控股及投资的公司有：盛世中皇控股有限公司、中皇汇通投资有限公司、集团旗下全资拥有建装业设计研究院、建装业体育产业发展有限公司、建装业新材料科技有限公司。集团与数十家大型企业组成战略联盟，具备强大的整合能力，高质量地完成了各类酒店、银行、学校、会所、会堂、机场、地铁、商场、医院、办公楼、住宅、别墅、厂房等大型装饰装修工程，呈现了强强联手的规模化运作优势。特别是在长期与金融业单位的合作中，总结出一套与金融业单位特性相适应的合作模式，使建装业成为国内国际金融业装饰领域的领军企业。

深圳市建装业集团一直秉承"诚信立业、以人为本、合作共赢、优质服务、开拓创新、铸造品牌"的经营理念！坚持"立足深圳、遍及全国、辐射全球"的经营方针！遵循"守信誉、重管理、重安全，建优质、鼎尚之工程，服务于社会各界"的企业使命！与时俱进，锐意进取，群策群力，携手共进，共同铸造"建装业"这一卓越品牌。

深圳市建装业集团愿与国内外各界人士在设计研究、建筑装饰、新材料研发、项目投资等领域展开广泛合作，共铸辉煌！

深圳市建装业集团施工建设中的海南三亚美丽之冠七星酒店

改善公共人居环境 助力城市繁荣

全球服务热线：4000 929 500

—— 文化·科技·节能·环保 ——

深圳市建装业集团股份有限公司
深圳市南山区深南大道西10168号佳嘉豪商务大厦3-9楼
电话：0755-26500008　传真：0755-26631238
邮箱：jzy@jzy.cn　网址：www.jzy.cn

人文之光　传世百年
Light of the humanities last forever

威伦斯
RENAISSANCE

威伦斯(Renaissance)——宝辉国际灯饰集团有限公司旗下品牌，一个设计灵感源于意大利文艺复兴艺术天堂佛罗伦萨的灯饰品牌。

威伦斯(Renaissance)背后，有着深厚的意大利文化底蕴，尤其是文艺复兴时期"人文主义"的思想浸润。古希腊的经典文化、意大利的浪漫、文艺复兴时的文化艺术精华，共同构筑了Renaissance的文化基因，打造经典的设计理念和艺术相融合，注定了产品风格的外在体现，同时也注定了Renaissance内在的思想张力。

集团总部： 香港九龙旺角亚皆老街8号朗豪坊办公大楼3715室
中国品牌营销中心： 广东省中山古镇新兴大道东路10号

威伦斯全国专卖店：

北京	天津	沈阳	长春	大连	大同	太原	丹东	石家庄
齐齐哈尔	赤峰	孝义	上海	宁波	杭州	无锡	常州	
烟台	南通	温州	台州	盐城	余姚	富阳	慈溪	昆山
湘湖	成都	昆明	重庆	郑州	武汉	西安	洛阳	荆州
贵阳	宜昌	沙市	广州	厦门	长沙	惠州	衡阳	

www.renaissance-dl.com

 香港寶輝國際燈飾集團·榮譽出品
THE WORLD'S LEADING LIGHTING SPECIALISTS

PERFECT STONE MODEL
FIVE STAR STONE KINGDOM
完美石材典范
打造五星级石材王国

雅高矿业(股票代码：03313.HK)广东独家经销商

HC STONE
皇朝石材

皇朝石材是最早实现石材集成、一站式采购的终端服务商，历经10多年的发展历程，现已完成矿山开采、加工、销售、设计、安装、维护为一体的综合运营，为我们的高品质客户提供最自然、最舒适、最奢华、最与众不同的选择，完美展现石材独有的原创和独特的装饰效果，缔造奢华夺目的艺术空间。

皇朝石材占地面积27000平方米，采用进口全自动电脑化石材加工设备，拥有高素质工程师、技术员、品质管理人员、生产员工共200多人和完整、科学的质量管理体系。

主要产品包括进口、国产的大理石，花岗岩板材和石线、圆柱、浮雕、拼花、马赛克等各种异型工艺制品。业务范围主要涵盖大型的楼盘、星级酒店、商城、地铁、办公楼、会所、别墅等。

皇朝石材倾力打造专业的营销及服务团队，赢得国内外经验丰富、观点独到的客户的青睐。做全球全面、个性化的石材供应商，为尊享级客户提供独特而全面的高品质产品及安装服务，打造中国一流的石材集成化企业。

皇朝石材标志工程：深圳T3航站楼四层安检大厅，深圳T3航站楼平安银行VIP接待厅，卓越时代广场，深圳罗湖边检出入境大厦，中国银行蛇口鹏龙支行，工商银行宝安丽景支行，平安银行景田支行，佳兆业大都会可域酒店，深圳回品酒店，广西北海富丽华大酒店，山西太原铂尔曼酒店，湖南桃江华美达酒店，金积嘉集团万国食品城，华南城总部大楼，华南城发展中心，华南城东北城总部，保利地产广州琶洲项目，招商地产伍兹公寓，万科双月湾，深圳南岭一半山，新世界地产世纪御园，湖北宜昌恒信中央公园，沈阳紫薇仙庄，安徽徽商集团酒楼项目，海底捞广州店，海底捞深圳店及深圳高端私人别墅等。

尊贵服务热线：
13823237666 / 13026626666

香港皇朝石材集团控股有限公司
广东皇朝石材工艺有限公司
深圳市皇朝石材有限公司

深圳市龙岗区南湾布澜大道盛宝路皇朝产业园
Tel.86 755 8996 3888 Fax. 86 755 8951 2555
www.huangchaostone.com